Vehicle Multiplex Communication

Serial Data Networking Applied to Vehicular Engineering

Other SAE titles of interest:

IDB-C Data Bus:
Report on Studies for Modeling, Simulation, and Signal Analysis; and EMC/EMI Measurements and Testing
by Daniel J. Guzman
(Order No. RR-014)

In-Vehicle Networks, Safety Critical Systems, Accelerated Testing, and Reliability
(Order No. SP-1783)

Software/Hardware and Systems Engineering
(Order No. SP-1781)

Vehicle Multiplex Communication

Serial Data Networking Applied to Vehicular Engineering

Christopher Albert Lupini

SAE International™
Warrendale, Pa.

For permission and licensing requests, contact:

SAE Permissions
400 Commonwealth Drive
Warrendale, PA 15096-0001 USA
E-mail: permissions@sae.org
Tel: 724-772-4028
Fax: 724-772-4891

Library of Congress Cataloging-in-Publication Data

Lupini, Christopher Albert.
Vehicle multiplex communication : serial data networking applied to vehicular engineering / Christopher Albert Lupini.
p. cm.
Includes bibliographical references and index.
ISBN 0-7680-1218-X
1. Automobiles—Electronic equipment—Computer networks. 2. Automotive computers. 3. Computer interfaces. 4. Data communication systems. 5. Multiplexing. I. Title.

TL272.53.L87 2004
629.2'73—dc22 2004042962

SAE
400 Commonwealth Drive
Warrendale, PA 15096-0001 USA
E-mail: CustomerService@sae.org
Tel: 877-606-7323 (inside USA and Canada)
724-776-4970 (outside USA)
Fax: 724-776-1615

ISBN 0-7680-1218-X
SAE Order No. R-340
Printed in the United States of America.

Acknowledgments

The author thanks Delphi Electronics and Safety for assisting him in writing this book. Appreciation is also given to the many reviewers of the manuscript, especially Matt Ruff, Nigel Allison, and Tim Thomas. Also, a special thank you to my wife Carrie for her unlimited support during this long process.

Table of Contents

Chapter 1

Introduction

Overview

This text is an introductory overview of vehicle multiplexing. The intended audience is engineers and managers working in the field of automotive and other vehicular electronics. The content for this book was developed from a three-day class on in-vehicle networking that has been taught in the public sector since 1993. A series of classes covering this material, and much more, are available now through Dearborn Group Technology, Inc. (www.dgtech.com).

Multiplexing, or in-vehicle networking, is one of the most popular areas of vehicle electronics currently. All vehicle original equipment manufacturers (OEMs), electronic control unit (ECU) suppliers, and integrated circuit (IC) vendors to the automotive industry are working on multiplex products. Multiplexing is the technology of applying serial data communications to the electric/electronic functions of a vehicle.

Multiplexing can be best discussed at three levels: the vehicle level, the ECU or component level, and the IC level. Within each level are partitions for software and hardware, and within each partition are divisions of functionality such as buffer size.

The following chapters will help the reader to acquire a basic understanding of vehicle multiplexing systems primarily from the viewpoint of the passenger car and light truck. Some discussion of heavy-duty and off-road vehicle multiplexing will be presented, as will be a look at industrial automation—a fast-growing multiplex field already eclipsing automotive usage. Hereafter, "automotive" will be used interchangeably with "vehicle."

Hardware and software issues will be presented for the more common protocols available. This survey should enable the reader to understand how multiplexing is generally done, so new protocols should present no difficulty as they arise. Much like software languages, all multiplex protocols have many things in common, and new schemes are simple to learn, once one or more are mastered.

Although this is an introductory and background text, information is presented on important technical issues such as electromagnetic compatibility (EMC). Within this realm, problems with electrostatic discharge (ESD), radiated emissions, and radiated immunity are important topics.

It is assumed that the reader has education and/or experience in electrical engineering. Some background in microcomputer operation and programming is also helpful. Basic techniques of binary/hexadecimal data conversion and computer architectures will not be presented.

The Contents of the Book

The remainder of this book includes the following chapters:

Historical Perspective

- How and why did multiplexing come about in the first place?

Background

- Terms and definitions needed to discuss multiplexing

Onboard Diagnostics (OBD)

- What is it?

Encoding

- Pros/cons of various encoding techniques

Error Handling

- How to protect from disruptive transients?

Media Characteristics

- What do we actually use to connect the nodes?

The Vehicle Level

- A look at multiplexing in general from the viewpoint of the vehicle

The Electronic Control Unit (ECU) Level

- A look at multiplexing in general from the viewpoint of the component

The IC Level

- A look at multiplexing in general from the viewpoint of the IC

Electromagnetic Compatibility

- How does the multiplex system interact with the outside world?

What Is a Standard?

- Introduces the world governing bodies for standards

Class A Protocols

- The low-speed protocols

Class B Protocols

- The medium-speed protocols

Class C Protocols

- The high-speed protocols

Diagnostic Protocols, Airbag Protocols, X-by-Wire Protocols, Mobile Multimedia Protocols, and Wireless Protocols

Data Link Usage

- Protocol comparison chart
- Company-by-company look at who is using what

Manufacturing Automation

- Industrial usage of multiplexing

Industry Activities

- What are General Motors (GM), Ford, and DaimlerChrysler doing together?
- Status of any worldwide consortia

Future Trends

- What's next?

Two appendices are included that give common terminology (glossary) and acronyms used in this book. The last two appendices are a survey of some of the known automotive (or related) protocols and a list of relevant International Organization for Standardization (ISO) and Society of Automotive Engineers (SAE) committees.

Chapter 2

Historical Perspective

Introduction

Multiplexing development has been in vogue since the mid-1970s. Initially, it was seen as a way to ease the diagnosis of electrical functions in vehicles [1]. Multiplex usage in military applications (i.e., MIL-STD-1553) was already well established at that time.

The surge of computing power primarily due to the emissions requirements of the late 1970s and early 1980s created renewed interest in multiplexing and in-vehicle networks. By the mid-1980s, most vehicle manufacturers had at least one serial data link protocol in production.

The 1990s were supposed to be the "boom" years for multiplexing. Unfortunately, leveling-off of electronic content and cost restraints put a serious damper on new applications, and this boom has turned into a "fast burn."

The Multiplexing Paradox

The multiplexing paradox involves the apparent contradiction between the popularity of multiplexing and actual usage of the technology. Although it is a powerful automotive systems architectural tool, it has not gained the widespread usage once predicted.

These are some of the reasons that multiplexing has not caught on as it should have:

1. It is not cost-effective.

   This is true in many cases because the cost comparison has looked simply at removing several wires and connectors and replacing them with additional protocol handling circuitry. This is not a fair analysis because it overlooks the cost savings of easier problem diagnosis, simpler servicing, better future expandability, easier removal/addition of vehicle options, and more choices for function partitioning and distribution among the ECUs.

It is still, however, the most powerful argument against multiplexing. Unfortunately, saving a penny today is considered more worthwhile than saving a dollar several years down the road.

2. It breeds competition.

   Believe it or not, multiplexing is considered a threat to ECU suppliers because it opens the door to more competition to supply components with a common serial data link interface such as the controller area network (CAN). This is mostly in the U.S. industry. Europe's vehicle OEMs are almost entirely dependent on outside ECU suppliers, and these suppliers have always had a lot of competition.

3. Little or no perceived customer benefit.

   Customer means the vehicle consumer. "Multiplexing" is not listed as content on the vehicle window sticker. How can one make the vehicle consumer realize the difference between a car with multiplexing and one without? The key technical point is that almost anything multiplexing is doing could have been accomplished with discrete wires (if space were available).

These are some of the positive reasons cited for using multiplexing:

1. Decreases the size of wiring harnesses

   Multiplexing signals can definitely reduce the number of interconnecting wires in the vehicle and save weight and space. Where this has the most effect is in constrained areas such as door pillars and underhood.

2. Economies of scale

   With a standard multiplex protocol, there should be a reduced piece cost of the hardware (i.e., ICs) that is included in the nodes. The theory says there will be a large demand for chips to implement the common interface(s), and therefore, prices will fall. We are witnessing this today with CAN.

3. Sharing of parametric data

   This item has already been proven in production. Redundant sensors, such as those for vehicle speed, can be eliminated because one sensor can supply its parametric information across the network for use by multiple nodes.

4. Increased reliability and test

   Multiplexing ECUs should allow better diagnosing of their circuitry because they can transmit and receive data from any node via one access port, that is, the data link controller (DLC). A built-in test can be designed in because the nodes may communicate with each other and check each other's functions upon initialization.

   Including the benefit of number (3) above, the hope is that low-cost vehicle electronic diagnostic tools will become available to the general public. The dream is for a \$99 handheld scanner so that shade-tree mechanics may troubleshoot their vehicles. Now, some government entities may not like this because of the possibility of tampering with emission control functions, but most vehicle consumers would welcome a cheap tool.

   A powerful topic relating to diagnosis and test is the idea of having a flight recorder, or "black box," as part of the vehicle's electronics [2]. As with aircraft, this box's contents could help in recreating a crash. IEEE 1616 is an attempt to standardize this. More benignly, it could also be used to troubleshoot malfunctions—especially those that do not occur with a technician at the wheel.

5. Greater flexibility in vehicle content

   If the rule of open architecture is followed (see Chapter 3), multiplexing should greatly help the problem of creating "world cars" and other kinds of flexible platforms. Vehicle options could be added more easily during assembly, or even at the dealership—sometimes months or years after the original sale.

6. Lower assembly and installation costs

   This is an area that must be considered when comparing the cost/benefits of multiplexing to discrete wiring. Multiplexing should make it easier to design, assemble, and maintain the vehicle.

7. Adding the impossible function

   Finally, many people believe that probably the most important enabler for multiplexing is that it can help to include a feature or function that otherwise would not have been part of the vehicle.

   Examples of this are the flight recorder, a built-in test among the ECUs, a $99 scan tool, the mobile office, and so forth.

An Early Multiplex Design

Figure 2.1 illustrates one of the first multiplexed networks to be found in a production vehicle [3]. Note that this design also illustrates the common occurrence of multiple networks. The 160 b/s connection was strictly for off-board diagnosis. The 8192 b/s link was the universal asynchronous receiver/transmitter (UART) bus that connected three ECUs to the assembly line connector. A 1 kb/s Entertainment & Comfort (E&C) bus connected the cathode ray tube (CRT) controller to the remote radio chassis.

Vehicle Multiplex Comparison to Industry

Compared to other (nonautomotive) schemes, in-vehicle networking is rather slow and suitable only for short distances (Fig. 2.2). Automotive needs, however, are much more demanding than most other uses. The in-car environment is harsh with respect to temperature, vibration, and electrical noise, not to mention that cost has to be low in a business where pennies count.

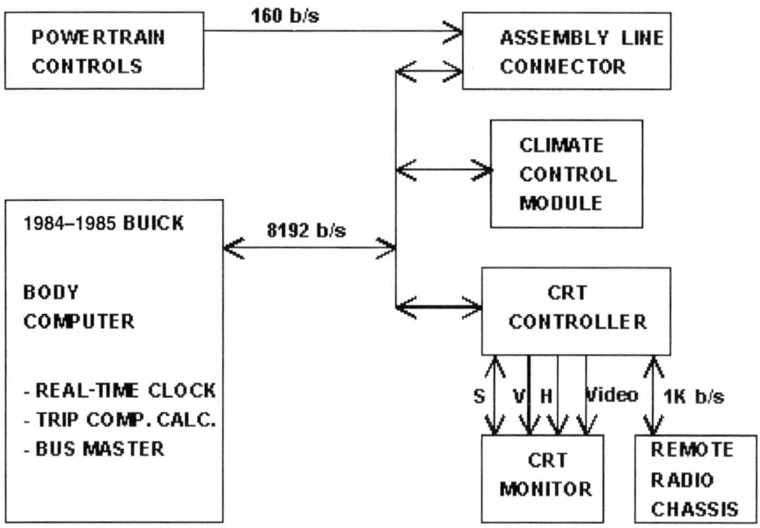

Fig. 2.1 Early multiplex implementation.

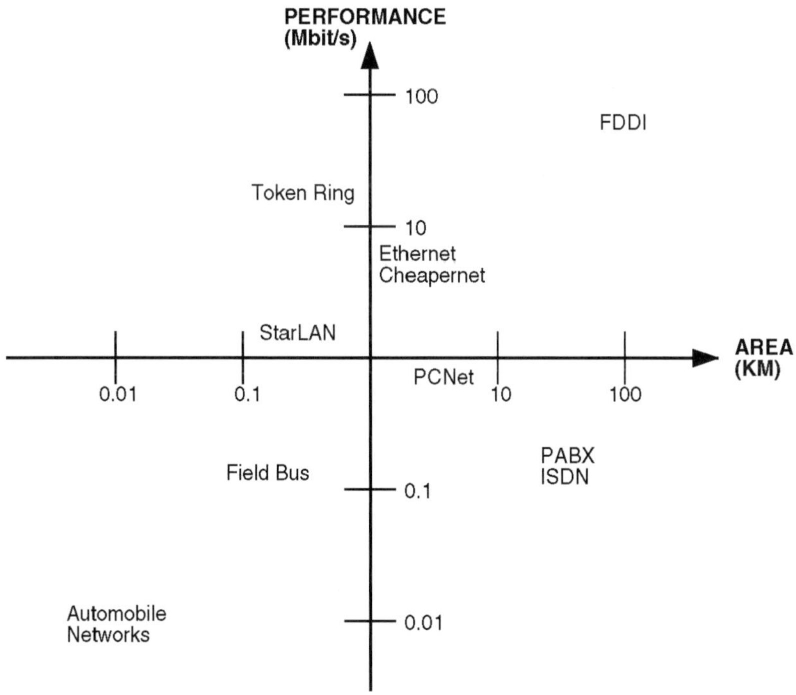

Fig. 2.2 Multiplex performance comparison.

As an example of the extent to which serial data links have been already used in automobiles, GM has used the following in production (sometimes several in the same vehicle):

- 160 bps diagnostic bit stream 1980
- 8192 bps UART 1985
- 1000 bps E&C 1987
- Allante multiplex bus 5000 bps 1987
- Class 2 10.4 kb/s 1994
- CAN (Europe only)1994

References

[1] Bell, R. "Multiplexing—Past, Present, and Future." SAE Paper No. 760178, Society of Automotive Engineers, Warrendale, PA, 1976.

[2] Buczko, G., and Hein, J. "The Need for a Flight Recorder Within a J1850 Networked Vehicle, and the Potential of Information Retrieved from Onboard Devices." SAE Paper No. 952580, Society of Automotive Engineers, Warrendale, PA, 1995.

[3] Christman, J.E. "Multiplexing for Allante." SAE Paper No. 870521, Society of Automotive Engineers, Warrendale, PA, 1987.

Chapter 3

Background

Introduction

This chapter presents the basic terms, definitions, and other foundation material necessary for discussing serial data link multiplexing. Also introduced are the idea of the ideal data link and the when, what, where, and why of multiplexing.

The Ideal Data Link

The ideal data link has the following attributes [1]:

- Open systems flexibility
- High reliability in a noisy environment
- Reduction of cost and manufacturing complexity
- Minimum central processor unit (CPU) burden for communication
- Maximum programmer transparency
- Guaranteed data consistency
- Short transmission times
- No added cost

The ideal data link does not exist! Therefore, we are forced to deal with a hodgepodge of standards. Most schemes are some compromise among these attributes. Keep in mind that not all of these items are important for all applications.

When Is Multiplexing?

As covered in Chapter 2, multiplexing is happening now. It has been around for a long time and will be around for a long time to come. One of the first automotive papers on vehicle multiplexing is referenced in [2]. This discussion dates to 1976 and states that "Multiplexing has recently become a subject of considerable interest in the automotive industry." It goes on to say that the concept is not new and continues with a brief history of multiplexing, outside the realm of automotive, to that point. An interesting statement from this paper is, "Perhaps, the time is at hand for its application in the automotive field." So here we find the beginning of the study of automotive multiplexing.

The paper "Multiplexing—Past, Present, and Future" by Robert E. Bell of Essex International accurately predicted where this technology was going and what it could do. Mr. Bell foresaw the dominance of time-division multiplex schemes in favor of multichannel amplitude modulation or frequency modulation. He also predicted, correctly, that multiplexing would have a hard time displacing most wiring in a vehicle because of the cost it adds.

Reference [3] has been another bellwether predictor of the trials and tribulations that vehicle communications would encounter. In this paper, the author estimates the cost of a time-shared multiplex system at 45% more than a conventionally wired system. This is precisely where most of the development, research, and engineering has been expended—how to use the benefits of multiplexing and minimize its weaknesses.

What Is Multiplexing?

Multiplexing is a way of transmitting digital signals between electronic components. As explained in Chapter 2, it has its origins in computer systems architecture. Multiplexing has been extensively applied to military and heavy-duty truck/construction equipment.

Multiplexing, also referred to as "in-vehicle networking," is often confused with other "information and communication" systems such as the cell phone, navigation, and so forth.

Where Is Multiplexing?

Also covered in Chapter 2, multiplexing can be found in all facets of automotive electronics. It also has a wide following in manufacturing automation and other industrial uses.

Why Multiplexing?

Much of the "why" of multiplexing can be found in the benefits section of "The Multiplexing Paradox" in Chapter 2. These are some of the common reasons for needing serial data links in automobiles:

1. Facilitation of Distributed Control

   Multiplexing was intended to support decentralized computing and loose coupling between the ECUs in a vehicle. It gives some partitioning choices for determining how much CPU horsepower and memory are necessary at each node. The possibility exists for very simple nodes, probably relying on a more powerful central node for support. An example is a master–slave arrangement.

2. Increase in Electronic Content

   This reason has been stated in numerous technical papers for years. Since the early 1980s, there has been an explosion of computing power within the automobile. The bulk of this can actually be traced to emissions requirements, but now we also have a growing safety awareness. This creates the need for many of the following new nodes on our networks:

   - Antilock brakes
   - Traction control
   - Heads-up display
   - Navigation
   - Drive-by-wire
   - Night vision
   - Mobile media

   Coupled with these new demands are the implied requirements for better performance and reliability. Multiplexing can help achieve these goals by providing less wiring that can be a target for electromagnetic interference (EMI), for example. Having a single-point connection to many ECUs at once also eases testing and diagnosing.

   An increase in electronic content will affect all models, not just high-level luxury vehicles. Eventually, options tend to filter down the line from

Cadillacs to Cavaliers, for example. At GM, the Corvette has been the top platform for introducing the latest technologies that then flow down to other models in later years.

3. Delay in Introducing Some Systems

   Large wiring harnesses create difficult vehicle design issues. This and other distributed computing partitioning choices can be alleviated by multiplexing and allow for quicker, smoother design.

   As an example of wiring size increase, the following "milestones" are given:

   - 1948 average car wiring: 10 lb, 150 ft, 35 connectors, 75 terminals, 55 wires
   - 1983 upscale Nissan 1.1-in. main wire harness at 48.5 lb
   - 1993 Corvette had three miles and 200 lb of wire (about 50 wires running in and out of the doors)
   - 1994 average car wiring: 65 lb, 5280 ft, 300 connectors, 2000 terminals, 1500 wires
   - 1995 upscale Nissan 1.8-in. main wire harness at 81.6 lb (cost Nissan approximately 0.3 mpg)

Multiplexing Is a Tool

A key point to keep in mind is that multiplexing is a tool to be used by vehicle electronic architects to facilitate vehicle-level electronic design. This has not been done to a large extent by anyone to date. It forces the vehicle platform architect to develop a distinct set of nodes and the required interconnecting network(s).

There is some fear within the electrical engineering community that too much responsibility is going to be put onto a data channel. In some cases, nodes that change the design parameters may be added and/or deleted after vehicle assembly. There is the possibility for a single-point failure to take down an entire network. Although this is true, it has to be balanced (risk assessment) against the positive attributes that are to be gained. Usually, multiplexing wins the reliability argument because wiring shorts and opens are very rare.

Popularity of Multiplexing

Multiplexing has enjoyed modest, if not exponential, growth. As an example, Table 3.1 indicates the phenomenal growth of multiplexing in Europe. The numbers indicate the multiplexing content (in percent) of that subsystem for a particular year.

A Standard Data Link?

A standardized data link is more desirable than a proprietary one. It is easier for the vehicle manufacturer to source ECUs and integrate them into the automobile. Common connectors may be used. This provides more choices

TABLE 3.1
FORECAST OF EUROPEAN MULTIPLEX USAGE

Application	1990	1995	2000	2005
Antilock Brake System (ABS)	17.5	38	55	60
Traction Control System (TCS)	0.2	2	8	18
Airbag	3	52	68	70
Multipoint injection (gasoline engines)	28	43	58	64
Single-point injection (gasoline engines)	15	33	25	23
Diesel injection (diesel engines)	14.2	20.5	29	31
Air conditioning	8.5	20	37	41
Automatic transmissions	5	17	30	36
Navigation systems	0	0.001	2	10
MULTIPLEX	**0.67**	**2.01**	**36.7**	**60**
Electrical/electronics content (in %)	11	17	25	28
Total value of electrical/electronics (in billion ECUs*)	7.5	13.2	22.4	26.4

Source: Hansen Report, November 1994

of ECUs, and the parts that make them are available. Of course, the nice thing about standards is that there are so many from which to choose. To make things simpler, we group protocols into classifications based on function and bit rate.

SAE Classifications

SAE has defined three classes of serial data communications: Class A, Class B, and Class C. It is an implied hierarchy with Class A at the bottom. Class B is a step up and can also do Class A tasks. Class C is at the top and can also accomplish Class B (and thus Class A) tasks. There is some talk that a Class D category may be needed in the future [4].

Class A:

1. Convenience operations
    - Entertainment modules, trip/mileage computers, and so forth
2. Low speed (<10 kb/s)
    - Maximum (fastest) message rate is probably around 100 ms (limit of human perception)
3. Intent: reduce cost, power, CPU throughput, and EMI
    - Inputs may be switches/sensors, and outputs may be power drivers/motor controllers

Class B:

1. Most information transfer
    - Engine, transmission, cluster data passing, and so forth
2. Medium speed (10 to 125 kb/s)
    - Maximum message rate is generally around 20 ms

3. Intent: legislated diagnostics, general purpose

Class C:

1. Real-time, or critical, control
    - Suspension damping, ABS, engine control, and so forth
2. High speed (125 kb/s to 1 Mb/s+)
    - Maximum possible message rate of < 5 ms
3. Intent: needs the intelligence and speed of a high-throughput system

Class D?:

1. *1 to 10* Mb/s *(for example, Ethernet?)*
2. *Year 2005 +*

The classes of communication carry a cost penalty with increased speed. As in car racing, "speed costs." There is typically an exponential increase in price also. However, the magnitude increase in bit rate is actually quite economical, as shown in Fig. 3.1. On a "cost per bit per second" basis, it would be very desirable to use a Class C network for most applications. This, however, is a long-range, farsighted approach not generally suitable for automotive electronic engineering.

We usually end up with multiple buses per vehicle—at least one for each class of communication. Chapters 13 through 20 discuss some of the more or less popular protocols used by automakers grouped by class, including some newer divisions such as airbag buses and mobile media.

As for critical or safety-related functions, there is a broad definition as to what is "critical." Certain functions are obviously critical to the operation of the vehicle—for example, airbag deployment, traction control, and so forth. Others are not so obvious. Some of these are seemingly minor but could distract the driver [5]. An example would be a driver's seat moving out of

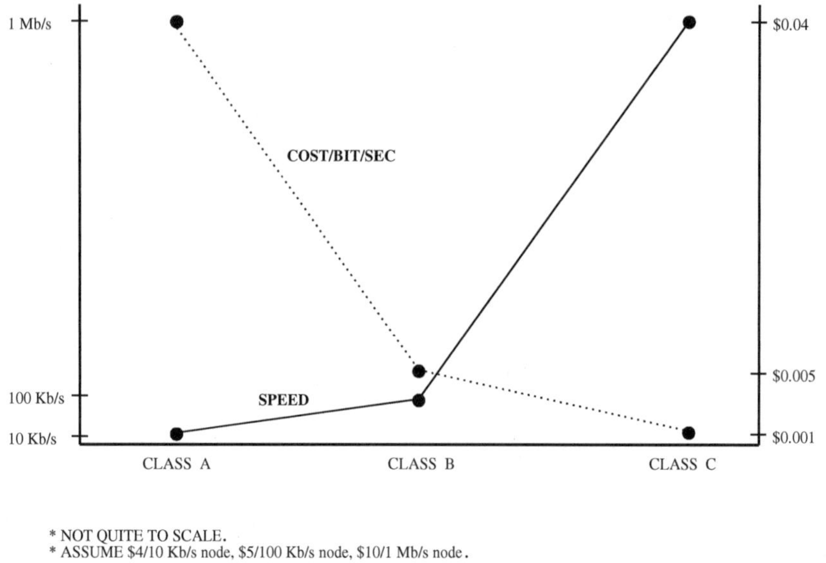

Fig. 3.1 Cost versus speed.

control toward the dashboard because of faulty bus handshaking. Drivers in other vehicles could be distracted or confused by improper signal operation due to bad serial communication. The result is that even Class A functions can often be classified as critical.

Definitions

Several terms need to be defined and understood to discuss multiplexing. Although many of them are listed in the glossary and acronym appendixes, some of the more important ones are discussed here.

Multiplexing

The word "multiplexing" means having multiple sources, and consumers of electronic data share a common data channel (Fig. 3.2).

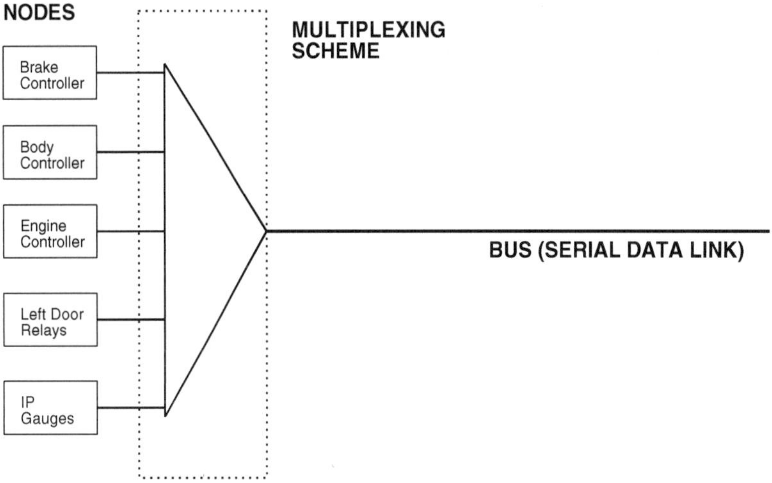

Fig. 3.2 Illustration of basic multiplexing.

Network

As seen in Fig. 3.3, a network is a collection of nodes connected in some sort of topology such as a bus (topologies are discussed later). A node may be an ECU such as an engine control module (ECM) or antilock brake system (ABS) controller. It also may be a sensor, actuator, and so forth. A node could also be another network if there is a gateway to another link. Nodes are tapped off the network via a stub. Usually, the stub is very short and can be found inside the node to eliminate splices in the wire harness. Stub length is critical for certain data link protocols to prevent ringing, overshoot, and so forth. The branching interval is sometimes very important from an ac/dc parametric viewpoint.

Intramodule Versus Intermodule Communication

Two other terms that need to be contrasted are intramodule communication and intermodule communication. Intramodule refers to data sharing inside a node. Examples are serial peripheral interface (SPI), serial communications interface (SCI), 8/16/32-bit parallel address/data interfaces, and so forth, between a CPU and peripheral ICs. Sometimes, an explicit clock signal is included (i.e., SPI), and short distances are always involved. Figure 3.4 shows an example of intramodule communication.

Intermodule communication, on the other hand, is data sharing between ECUs. This is always a serial channel and may involve relatively large distances (e.g., 40 meters). The network in Fig. 3.3 is an example.

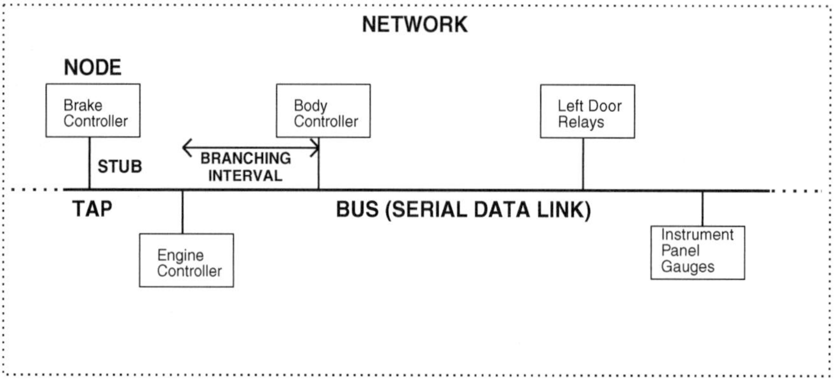

Fig. 3.3 Basic serial data network for definition purposes.

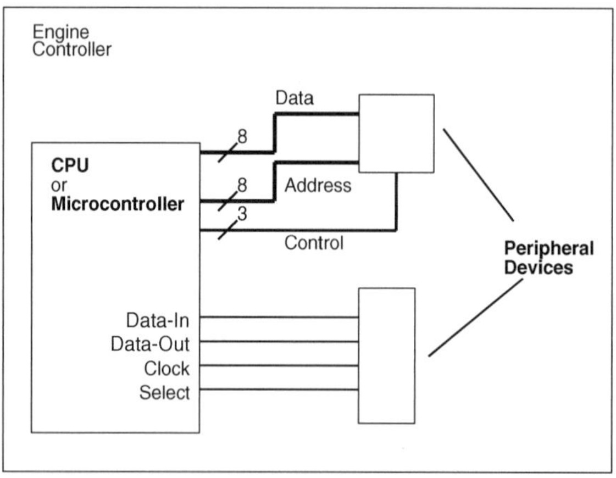

Fig. 3.4 Example of intramodule communication.

Intermodule communication, or multiplexing, is important because it has such a wide scope. It can involve all of the electronics in the vehicle. It can be used for special-purpose information passing between nodes such as a body computer and instrument panel CRT screen. Usually, it is used for extensive data and control sharing. Often it is useful for offboard communication with testers, scanners, memory programmers, and so forth.

Examples of Nodes

All of the following are possible nodes, not counting other possibilities such as sensors, actuators, relays, and switches.

- Engine
- Transmission
- Brakes (four controllers; one at each corner?)
- Suspension damping
- Radio
- Amplifier
- CD player
- Remote CD changer
- Power antenna
- Cellular phone
- Window switches
- Power mirrors
- Door locks
- Power-seat controls
- Tilt/telescope steering wheel
- Theft-deterrent system
- Horn
- Trunk lock
- Headlights
- Turn signals
- Courtesy lights
- Brake lights
- Trip computer
- Navigation system
- Instrument cluster

- Cruise control
- Electronic diagnostic system

Most of these can be found in production applications, and the list continues to grow. It usually does not make sense to replace simple one-wire functions with a serial data interface. However, functions such as horn and headlights could be activated by a theft-deterrent system by using a serial data link. Duplicate control/feed wires from switches and the theft-deterrent module could be eliminated.

Open Architecture

Another important concept is open architecture. An open architecture is very advantageous for automotive multiplexing. It means that nodes can be added or deleted without any modification in either hardware or software of any other nodes. Ideal openness is extremely hard to achieve. Even if no modifications are necessary at the other nodes, some provision is needed for adding/deleting nodes. Usually, this means that software has to be set up for the possibility of including or not including another node. As an example, the software in the ECM has to be able to react to an ABS module present or not present on the vehicle.

Broadcast

Broadcast, or multicast, means that when one node transmits a message, all other nodes (including the transmitter) see the message. Most automotive serial data links possess this attribute. It is useful because many times the data are functional and will be needed at more than one node.

Peer-to-Peer

Peer-to-peer communications implies a masterless link. With it, each node has approximately equal access to the channel. Usually, every node may begin transmission at the same time, and collisions are resolved by one of several methods (described later).

Baud Versus Bit Rate

The term "baud rate" is often misused and deserves some attention here. Many people equate baud rate with bits/second (b/s), but this is not a correct definition. Baud rate is defined as "n distinct signals or levels or transitions per second." It implies an information rate but does not define the exact units. Baud means "rate," so saying baud rate is redundant, anyway. Using the term baud rate instead of b/s can cause confusion. For example, pulse-width modulation (PWM) encoding transfers one bit of information with every bus period (two edges), whereas nonreturn to zero (NRZ) conveys two bits of information with every bus period. If one says a bus is running at 150 kbaud, does it mean 150 kb/s or 150 kHz? Hereafter, this book does not use the term "baud rate."

Protocol

A bus "protocol" is the set of rules that transmitting and receiving nodes must follow so that correct communications take place. The protocol usually specifies such things as bit rate, signal encoding, required error checks, acknowledgment methods, and so forth. Not all protocols include complete detail per the open systems interconnect (OSI) seven-layer model. Some protocols, such as J1939, specify most layers of the OSI model but are based on another protocol (CAN in this case).

Protocol Handler

A "protocol handler" is the usual term for the implementation of a protocol. It is typically a stand-alone IC or ICs and may consist at least partially of circuitry onboard a microcontroller. In some cases, the protocol handler may be 100% software as in bit-banging a simple UART bus. However, most bus protocols need dedicated hardware run by a CPU to accomplish their functions.

Protocol handlers are classified into three categories based on their data buffering capability. These categories are message level, byte level, and bit or symbol level. Basically, a message level handler deals with bus traffic on a complete message basis, whereas a byte level handler deals with bus content on a byte-by-byte basis. A bit or symbol level handler trades software complexity for hardware simplicity by forcing the CPU to perform bit manipulation to transmit and receive.

Gateway and Bridge

The term gateway is commonly used in automotive multiplexing to refer to a node that converts protocols between two different data buses. The correct term for such a function is actually "bridge," but gateway has become the de facto standard. For an example of an automotive bridge, see [6]. Implementing gateways and bridges (and also routers and repeaters) is not trivial. One must indicate which messages will flow across the boundary between networks, how often they will flow, to whom they are targeted, and so forth.

Acknowledgment

Acknowledgment is an important concept in data networking, and most protocols require some sort of acknowledgment scheme. A "positive" acknowledgment is some sort of active recognition of a transmitted message. An example of positive acknowledgment would be a receiver appending response bits or bytes to the original message (in-frame response). An in-frame response used for positive acknowledgment is considered a waste of bus bandwidth. The consumer of the message content has not possibly had time to act on the data before having to place an acknowledgment within the original message. Negative acknowledgment is usually the concept of "no news is good news." In this case, if no receiver complains about the message, the transmitting node assumes that it was OK. Another method of negative acknowledgment is to have receiver nodes respond back in some fashion if an error is detected. The error reported may be a problem with the physical network, message format, or message data, depending on the protocol.

Bandwidth

Figure 3.5 shows a rather obvious relationship between the required bus bandwidth and the level of bus function [7]. A powertrain node will have more bus load because of the message count and will require more bus bandwidth. A key point here is that powertrain control and lighting perhaps do not belong on the same bus. If all that is going on is general information passing and diagnostics,

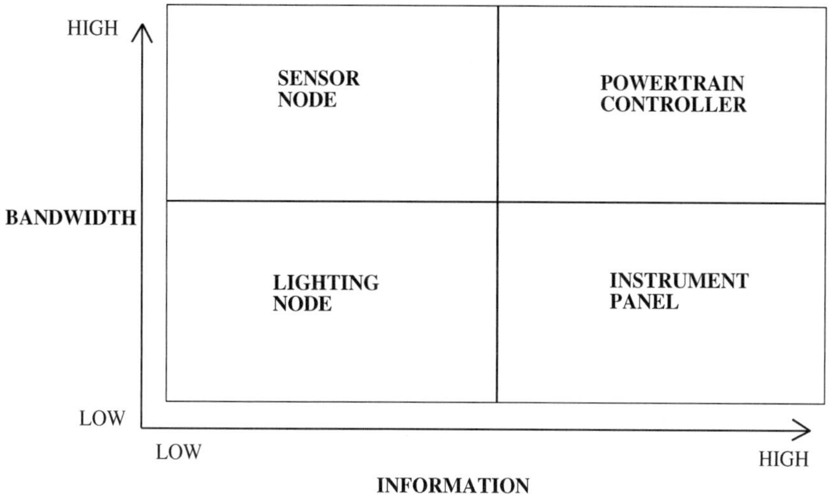

Fig. 3.5 Bus bandwidth versus control function.

it might work. But in general, real-time control functions do not belong on the same bus as lighting, seat controllers, and so forth.

Communication Direction

Several more definitions are in order. Most automotive bus protocols are half-duplex. This means that communication can take place in one direction at a time. On a half duplex bus, every node is either a transmitter or a receiver but not both at the same time. One distinction about the last sentence needs to be made—a transmitter usually receives its own messages. This does not make the communication full duplex, however, because the transmitter is not receiving data different from that it is transmitting.

Asynchronous Versus Synchronous Communication

Asynchronous and synchronous are terms to define the type of bus access, or time division, that takes place among the nodes. Asynchronous means in the absence of a common clock. There is no predefined timing between messages, and the nodes do not know when to expect messages. Transmitters provide their own clocking, and receivers recover bit timing off edges (voltage change, light on/off, etc.). The typical timing of asynchronous bus messages (frames) is shown in Fig. 3.6. The time T may vary between upper and lower bounds but is unpredictable for any given message.

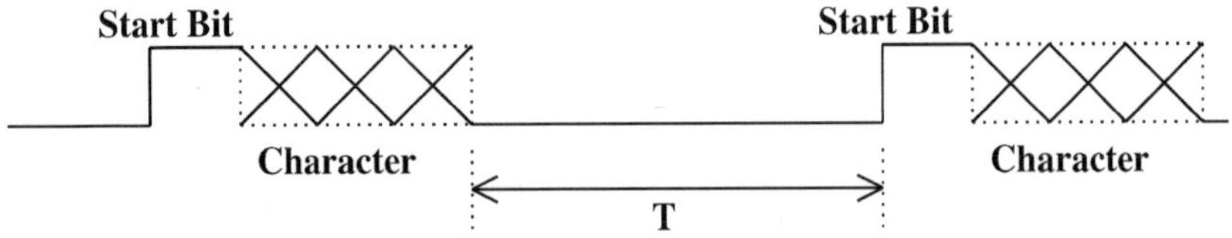

Fig. 3.6 Timing of an asynchronous message.

In asynchronous communications, the transmitting node determines when to transmit the "start" bit or symbol. The message bits to follow have some predefined timing, and usually, an error check (parity, checksum, cyclic redundancy check, etc.) is appended at the end. Receivers detect the arrival of the "start" character and will either sample (Fig. 3.7) or count to determine whether a bit is logically a zero or a one.

In asynchronous schemes, some form of resynchronization is needed. Usually this is done on certain edges within the message, so that the receiver will not "drift" and hence not know any longer where the bit boundaries are.

Synchronous, compared to asynchronous, timing is not usually used in automotive multiplexing. This method is most often used in intramodule data communication such as on a circuit board. The serial peripheral interface (SPI) bus is a well-known example of intramodule synchronous communications (Fig. 3.8). This type of communication has a known relationship between transmitters and receivers via a distinct clock signal. It is not desirable to run clock signals around a vehicle because of concerns about radiated emissions.

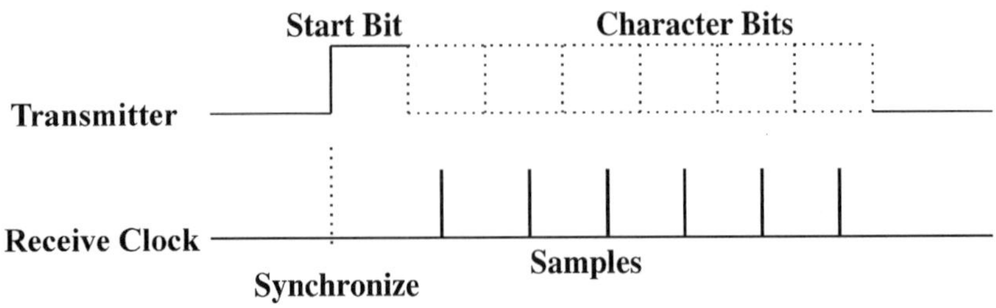

Fig. 3.7 Sampling an asynchronous message.

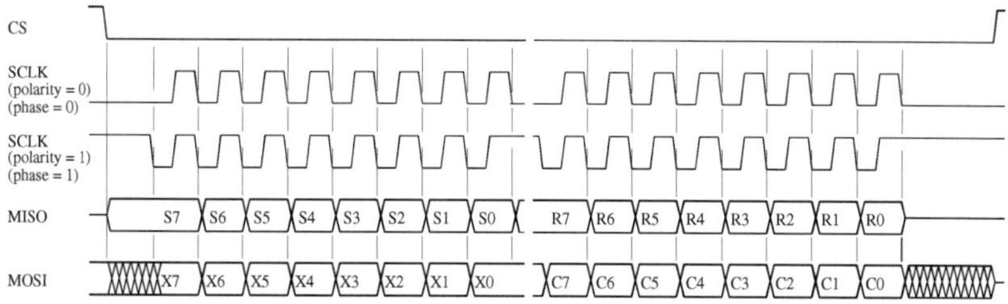

MISO is serial data output from peripheral
MOSI is serial data input to peripheral
SCLK is the SPI clock
CS* is the chip select

Fig. 3.8 SPI communication.

Besides asynchronous and synchronous communication, there are other methods of coordinating message traffic between nodes. One of these is pleisiochronous, which is "semisynchronous," probably using two different clock sources. Another is "isochronous," or "time-triggered." Time-triggering becomes important in dealing with safety-critical applications where messages must arrive at preset times. The main component of a time-triggered architecture is a precise global time clock among the nodes.

The following is a brief summary of these communication types:

1. Asynchronous

   - An event that can happen at any time
   - "Event-driven" data
   - CAN is an asynchronous protocol

2. Synchronous

   - An event determined by a predicted trigger condition
   - Data that are announced beforehand
   - SPI is a synchronous protocol

3. Pleisiochronous

   - Semisynchronous
   - Probably two different clock sources
   - Pleisiochronous Digital Hierarchy (PDH) is a pleisiochronous protocol

4. Isochronous

   - An event that must always happen at the same time
   - "Time-triggered" data
   - TTP is an isochronous protocol

Channel Sharing

Of the possible methods for dividing up the channel for multiple-node use, frequency division and time division, time division is the method of choice. The primary drawback of frequency division multiplexing is cost. In this method, the signal frequency is divided into subchannels, as with cable TV. Guard bands are required between subchannels, and each subchannel is assigned to one user. This permits full duplex communication between nodes but possibly wastes subchannels that are unassigned. A two-subchannel example is shown in Fig. 3.9.

IBM developed a commercial example of a wavelength division multiplex system for commercial voice/data communications. This system employs a single optical fiber with 25,000 GHz of theoretical capacity. This channel is divided into 10 bidirectional data streams using 20 wavelengths. These channels are further divided into 10,000 1-Gb/s subchannels spaced 120 GHz apart. This gives the equivalent capacity of 10,000 uncompressed High-Definition Digital Television (HDTV) channels. The system is protocol-independent and costs approximately $250,000 for a two-node basic implementation with a maximum spacing of 50 km.

Time division multiplexing is the usual method of channel allocation for automotive use. Multiple users share the single channel. Each user gets full use of the channel for a fixed or variable length of time. This brings up an interesting point with regard to the commonly used statistic, "bus utilization." Sometimes, the amount of network bandwidth used is described as 37% or 80%,

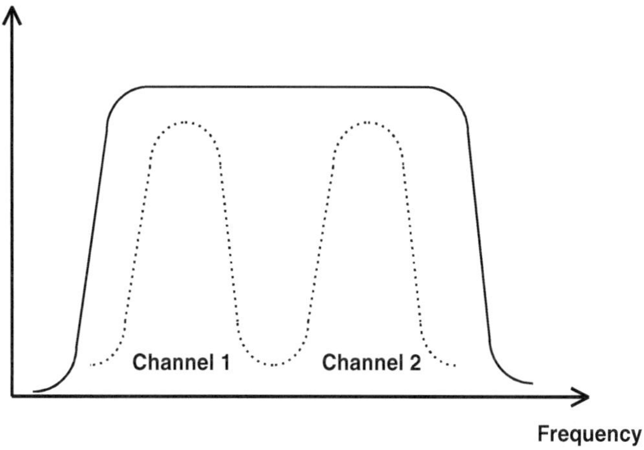

Fig. 3.9 Example of two-channel wavelength division multiplexing.

for example. This means that a time division multiplexed network is using 37% or 80% of the bandwidth. In reality, this should be stated as the bus being used 100% for 37% of the time. The bus is either busy (used), or it is not.

Bus Access

Now that we have determined the method of allocating the bus channel (time division), the next thing to define is the method of bus access. Bus access, or the method of sharing, takes one of three main forms: master–slave, round-robin (also called token-slot or token-ring), or arbitration (contention).

With master–slave, one node (master) has primary control of communication for the other (slave) nodes. The master gives the slaves access on some predetermined basis. A slave cannot transmit unless told to do so by the master. The main benefit of this approach is a fixed, known message latency. The primary disadvantage is that it does not provide for an open architecture. The master node must know which slave nodes are on the network and what their needs are. Slaves may be polled to determine if they need to transmit, or they may be given transmit opportunities at regular intervals. Probably the most important advantage of master–slave networks is the fixed latency, but there are other good attributes. These include easier implementation and generally lower cost because of decreased software.

Round-robin, token-slot, or token-ring are the names given to a method of bus access that does not have an explicit master node but is not a colliding (contention) bus. In this scheme, each node is given, *a priori*, one or more time "slots" in which it has use of the bus. The node allowed to transmit is identified by the content of a special message called a token. After this node transmits, it updates the identifier portion of the token and sends it back out. Like master–slave, this is not an open architecture, but its message latency is fixed.

The most interesting, and widely used, bus access method is arbitration, or contention. Colliding or arbitrating types of bus protocols belong to a generic family of access methods called CSMA/CD and CSMA/CR. The first two initials CS stand for carrier sense; this means that all nodes monitor the bus for activity and can tell whether or not it is busy. MA stands for multiple access and implies that all nodes can access the bus at the same time. CD means collision detect and is used by nonautomotive protocols such as IEEE 802.3 (Ethernet). There are collisions but no arbitration. If two or more nodes begin transmission at the same time, they will detect a collision and turn off their transmitters for a random length of time before trying again. CR or collision resolution takes CD a step further by providing a mechanism to resolve bus contention. This mechanism, called arbitration, was patented in 1975. Multiple transmitters resolve control of the bus channel via message priority bits. Simultaneous transmission continues until the bit desiring to transmit differs between the nodes. Nodes transmitting the "recessive" polarity of a bit (usually "1") stop transmitting immediately and remain receivers for the rest of the message. Contention is a "wired-or" of all transmitters, and arbitration resolution is a "wired-and" of all transmitted logic levels (for a zero-dominant bus). The primary benefit of arbitration is that no time is lost and one message always gets through.

A perceived disadvantage of a contention protocol is the possibility of long message latencies. Great care must be used in determining the priority of the message via its included header content. Very high priority messages will get through in no less than the time for the current message to finish. Obviously, very few messages should have the highest priority. Most messages will have some intermediate priority and, depending on bus traffic, have a variable latency. Some vehicle electrical architects do not like this attribute and will choose a master–slave or token ring type of protocol instead.

An important point to remember is that arbitrated types of networks have a best case message latency that is shorter than any other type of access method. Unfortunately, the other types will have better average and worst case latencies, and, their worst case latencies are fixed and thus predetermined.

So, the key to using a contention type of bus protocol successfully is to assign message priorities intelligently early in the network design process. Simulation tools are available to analyze the latency performance of the link before it is built.

Baseband Versus Broadband

One further clarification to be made in the introduction to multiplexing is the observation that all automotive bus protocols use baseband signaling as opposed to broadband. In baseband signaling, the information is placed directly onto the media without a carrier or intermediate frequencies, and so forth. The digital signal is modulated directly onto the wire(s) or fiber as ones and zeros.

An obvious example of broadband signaling is FM radio, where the information is transferred via an analog modulated carrier wave. This method is not used in automotive multiplexing because of the cost, complexity, and short distances involved.

References

[1] Phail, F., and Arnett, D. "In-Vehicle Networking—Serial Communication Requirements and Directions." SAE Paper No. 860390, Society of Automotive Engineers, Warrendale, PA, 1986.

[2] Bell, R. "Multiplexing—Past, Present, and Future." SAE Paper No. 760178, Society of Automotive Engineers, Warrendale, PA, 1976.

[3] Nevett, L. "Time-Shared Multiplexing System Applied to Motor Vehicles." SAE Paper No. 760181, Society of Automotive Engineers, Warrendale, PA, 1976.

[4] Allison, N. "Rethinking Multiplex." SAE Paper No. 941650, Society of Automotive Engineers, Warrendale, PA, 1994.

[5] Ward, D. "Integrity of Automotive Electronic Systems: A View from Europe." SAE Paper No. 940371, Society of Automotive Engineers, Warrendale, PA, 1994.

[6] Szydlowski, C. "A Bridge Linking CAN and J1850." SAE Paper No. 931808, Society of Automotive Engineers, Warrendale, PA, 1993.

[7] Eisenhart, S., and Baum, D. "Hardware/Software Partitioning for Automotive Multiplex Systems." SAE Paper No. 910711, Society of Automotive Engineers, Warrendale, PA, 1991.

Chapter 4

Onboard Diagnostics (OBD)

Introduction

This chapter takes a look at diagnostics, a major driver for automotive serial data links. Both legislated (emissions) and manufacturer-specific (service) diagnostics demand powerful communications strategies. The bulk of the discussion centers on diagnostic requirements related to pollution, or emissions.

Onboard Diagnostics (OBD)

"Diagnostics" generally refers to a broad category of functions within the automobile and between the automobile and outside world. Onboard diagnostics (OBD or OBD-I) refer particularly to government-mandated, emissions-related requirements [1].

GM, for example, has instituted emissions-related diagnostics (i.e., OBD) since the 1980 model year (MY) (Fig. 4.1).

Original official federal OBD requirements were vague, and the strategies to implement them were generally crude [2].

OBD-II

In the late 1980s, the California Air Resources Board (CARB) began action to implement a new, more stringent set of emission requirements for all passenger cars sold in California beginning in MY94. These new regulations would be called On-Board Diagnostics-II, or OBD-II, officially called: *Title 13, California Code of Regulations, Section 1968.1 "Malfunction and Diagnostic System Requirements 1994 and Subsequent Model Year*

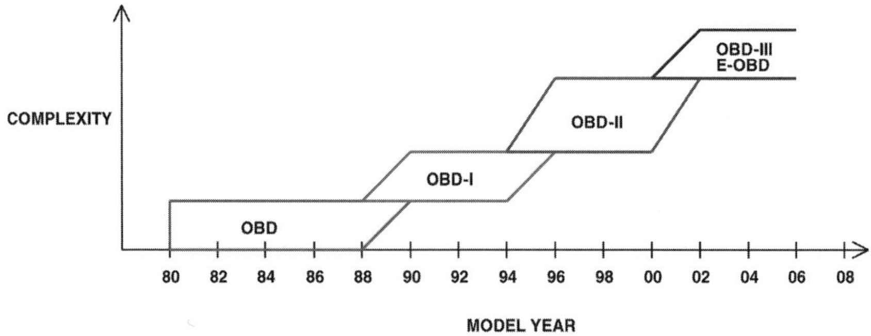

Fig. 4.1 OBD time table.

Passenger Cars, Light Duty Trucks, and Medium Duty Vehicles with Feedback Control Systems." A key feature of OBD-II would be that it would require a standardized serial data link on each vehicle, regardless of manufacturer, for interfacing an offboard handheld scan tool. This link would be used to communicate requests from the scan tool for certain emission data stored in the vehicle and to communicate responses. An example would be the number of misfires per mile and which cylinder(s) are misfiring.

CARB began soliciting ideas for the standard data link in 1988 and looked closely at SAE J1850. A little background on J1850—this was an effort by SAE to standardize a Class B protocol for in-car use. Unfortunately, GM proposed its new Class 2 protocol, Ford offered a PWM-based solution, and DaimlerChrysler its UART-based DaimlerChrysler Collision Detection (CCD) bus. SAE turned down the CCD offering but, unfortunately, accepted both the GM and Ford protocols. The data rates were modified to accommodate standard microprocessor crystal clock rates. For example, Class 2 was originally at 10 kb/s but was changed to 10.4 kb/s (a 96-μs average bit time). Ford's system became the standard corporate protocol (SCP) internal to Ford and runs at 41.6 kb/s.

The final choices that CARB allowed for OBD-II were J1850 and ISO 9141-2. Now, Keyword Protocol 2000 (KWP 2000) is also accepted as a legal interface because KWP 2000 is based on ISO 9141, which is widely used around the world.

The OBD-II regulations required vehicle manufacturers to install systems on all 1994+ light-duty vehicles (LDVs) and light-duty trucks (LDTs) to identify emission-related component deterioration accurately during the life of the vehicle. This was different from the old rules that required detection of only gross failures of components in the emission control system. Provision was made for new diagnostic trouble codes (DTCs) and a new scan tool for use by the emission test stations in California and elsewhere.

Whereas OBD-I targeted only malfunctions that would cause a drivability problem, OBD-II required monitoring for things that would cause an increase in emissions. The cost per vehicle was estimated at $78. After figuring in increased fuel economy, the final cost for the consumer was estimated at $36.

An additional benefit from OBD-II that its implementers desired was an increase in the security required to change the engine operating parameter calibration. In the past, most vehicle manufacturers provided an engine control computer with a removable programmable read-only memory (PROM) containing the engine operating parameter references. That memory chip contained data, such as base idle speed, basic timing advance, advance under various loads, and so forth. OBD-II was intended to use Flash memory in an unremovable IC. Under OBD-II, this memory can be programmed or reprogrammed only by a special tool that contains the security codes and electronic "keys" to access the calibrations. Although only a small number of vehicle calibrations were ever tampered with under OBD-I, it was considered a victory for pollution-control advocates that OBD-II would severely restrict such actions.

OBD-II did not apply to vehicles containing diesel engines. It also did not apply to heavy-duty trucks (more than 8600-lb gross vehicle weight), construction equipment, and farm machinery [3].

CARB Requirements

The CARB requirements were centered on a set of request and response messages. These "modes" had to be supported by vehicle manufacturers and test tools. The main definition of these modes is set forth in SAE J1979.

The following are some of the modes defined in J1979:

- Request Current Powertrain Diagnostic Data (Mode $01)
- Request Powertrain Freeze Frame Data (Mode $02)
- Request Emission-Related Powertrain Diagnostic Trouble Codes (Mode $03)
- Clear/Reset Emission-Related Diagnostic Information (Mode $04)
- Request Oxygen Sensor Monitoring Test Results (Mode $05)
- Request Onboard Monitoring Test Results for Noncontinuously Monitored Systems (Mode $06)
- Request Onboard Monitoring Test Results for Continuously Monitored Systems (Mode $07)

Additional, nonemission-related modes were defined in J2190 and were called "enhanced E/E Diagnostic Test Modes." These were above and beyond J1979 and were voluntarily implemented by vehicle manufacturers. The following mode ranges are defined by J2190:

- Modes $10–$3F defined for particular operations
- Modes $80–$9F reserved
- Modes $A0–$BF manufacturer specific

Besides monitoring catalytic converter and oxygen sensor(s) operation, OBD-II requirements included detecting engine misfire and codes for which cylinder(s) misfire. Misfire information had to include the number of misfires per mile, which is no easy task to determine. Leakage in the vapor recovery and purge systems had to be monitored. On OBD-II-equipped vehicles, a DTC is set if the gas cap is loose or left off! All other emission components were monitored for deterioration.

The following are the amounts of increase in emitted pollutants that had to be detected:

- 0.2 g/mi hydrocarbons (HC)
- 1.7 g/mi carbon monoxide (CO)
- 0.5 g/mi nitrogen oxides (NO_X)

The increased amount of software to accomplish OBD-II was a huge undertaking. General Motors reported in March of 1997 that a certain application took 60,000 lines of code and 15,000 calibrations just to do OBD-II functions. This was as much software as all other functions of the vehicle combined.

As the reader may have detected already, there were a number of documents that were required or related to an OBD-II electronic system. For example, to implement a "J1850 system" took a lot more than simply the J1850 document as a guide. More specifications were needed to "fill out" the definition. There also were many related documents that specified nonemission recommended practices and requirements for the test tools, and so forth. The pertinent documents listed by functional area are shown in Fig. 4.2. Boldfaced type indicates that it is a required part of an OBD-II system.

SAE publication HS-3000 contains many of these documents (not ISO 9141-2, J2205, J2008, J2300, or J2201).

Onboard Monitoring

OBD-II requires monitoring the following items once per vehicle trip:

- Catalyst efficiency (conversion efficiency)
- Heated catalyst (time to temperature)
- Evaporative system (airflow and leaks)
- Air injection system (proper functioning)
- Oxygen sensor (response time)
- Oxygen sensor heater (current or voltage drop)
- Exhaust gas recirculation (EGR) system (flow rate)

Protocol	**J1850** J1699		**ISO 9141-2**
Diagnostics		**J1979** J2190 J2205	
Messages		**J2178**	
Trouble Codes		**J2012**	
Security		**J2186** J2008	
Scan Tool		**J1978** J2300	
Scan Tool Interface		**J2201**	
Connector		**J1962**	

Fig. 4.2 SAE documents related to OBD-II.

These are items that must be continuously monitored:

- Misfire detection (percent misfire)
- Fuel system (ability to control fuel delivery)
- Other comprehensive component monitoring

Surprisingly, if a problem is detected, what to do about it is left mostly to the judgment of the vehicle manufacturer. The apparent severity of the problem will determine whether the malfunction illumination lamp (MIL) will light immediately, wait for more occurrences, or just log it as a DTC.

A couple of unexpected benefits from increased regulation such as OBD-II are a result of the many operating parameters that are now logged within the electronic system. One of these benefits is a "flight recorder" type mode of operation that can help repair shops troubleshoot problems with the vehicle, especially problems in the engine.

Another benefit is that OBD may allow a built-in test (BIT) in the future [4], which is the ability of onboard electronics to "diagnose itself."

Although originating in California, several other states have adopted OBD-II regulations. The U.S. Environmental Protection Agency (EPA) has accepted the OBD-II rules in all states through the 1998 model year. Thus, all vehicle manufacturers have chosen to implement OBD-II as federal emissions, not just for California vehicles.

Other countries are generally lagging behind the United States in emissions regulations. For example, Canada still uses a set of rules on a par with the U.S. OBD-I specs.

Europe is in about the same situation as Canada, except that an update is in development. The United Kingdom Department of Transportation proposed OBD-II as the basis for Motor Vehicle Emission Group regulations beginning in MY2000. These rules are called "Europe OBD" or "E-OBD" and would harmonize with "OBD-III" in the United States.

These new requirements would be contained in ISO, instead of SAE standards, and would contain requirements for connectors, communication protocols, trouble codes, and so forth. It is highly likely that they will be similar, if not identical, to the U.S. standards because of harmonization efforts between SAE and ISO. One area of likely difference will be in the indicator lamp standard where the existing ISO standard will be retained. The ISO task force working on converting SAE "J" documents to ISO documents is ISO/TC22/SC3/WG1/TF5.

As for communication protocol required standards, E-OBD is likely to be based on Keyword 2000. This high-level protocol may then be used on top of one of several lower-level protocols such as J1850, CAN, or ISO 9141. ISO 9141 has been used for diagnostics in European vehicles for many years and will continue to do so, including new emissions requirements.

It is predicted that these are the levels of pollutant increase to be detected by E-OBD:

- 0.4 g/mi HC
- 3.2 g/mi CO
- 0.6 g/mi NO_x

As in the United States, diesel engines have different requirements as do heavy-duty and off-road vehicles.

Onboard Diagnostics Concerns

Several concerns have been raised about ever-increasing OBD requirements. One of these is a side effect of fuel quality that may be bothered by required additives. These additives may cause false MIL illumination. The by-product sulfur can cause early MIL activation for catalyst and oxygen sensor operation.

Another concern is the extensive training that will be required to enable technicians to diagnose problems on OBD-equipped vehicles [5]. The OBD systems will be beneficial only if what they are saying can be properly interpreted by a service person.

The vehicle driver will also be affected by OBD. There are more instances where the "service engine soon" light will come on. The driver will "lose faith" if the MIL lights up often or sporadically. One example of this is the MIL illuminating when the gas cap is off or loose. There is no effect on the drivability of the vehicle, so the driver assumes that the light is malfunctioning. After a while, the light will be ignored.

Lack of world standardization in emissions regulations, service procedures, and general diagnostic tests is causing proliferation of techniques and rules. This will come to a head with increasing instances when vehicle manufacturers produce what they call "world cars." These vehicles are meant to be sold all over the globe and must conform to each country's requirements.

The Future

"OBD-III" requirements will roll out in MY04 and be gradually phased in through MY07. By MY07, the only acceptable offboard communication protocol for emission testing (at least in the United States and Europe) will be ISO 11765 CAN.

References

[1] Walsh, M. "A Comprehensive State Motor Vehicle Pollution Control Strategy for the 1990s and Beyond Part I: Light Duty Vehicles." SAE Paper No. 921056, Society of Automotive Engineers, Warrendale, PA, 1992.

[2] Bryant, P. "A Review of the Potential for Vehicle On-Board Diagnostic Safety Systems." SAE Paper No. 921596, Society of Automotive Engineers, Warrendale, PA, 1992.

[3] Stepper, M., Butler, S., and Zhu, G. "On-Board Diagnostics, A Heavy Duty Perspective." SAE Paper No. 951947, Society of Automotive Engineers, Warrendale, PA, 1995.

[4] Armstrong, L. "The Role of On-Board Computers and Networks for Fault Detection and Isolation." SAE Paper No. 902216, Society of Automotive Engineers, Warrendale, PA, 1990.

[5] Lindren, M., and Persson, P. "The Multiplexed Vehicle—A Maintenance Time Bomb?" SAE Paper No. 960391, Society of Automotive Engineers, Warrendale, PA, 1996.

Chapter 5

Encoding

Introduction

Encoding is an important topic for multiplexing and so is treated in its own chapter here.

What Is Encoding?

A digital signal must be coded onto a medium. Receivers must know how to interpret the signal to extract the data. Usually the serial data link protocol specifies the encoding to be used. Encoding is defining what conditions (voltage, current, etc.) constitute a particular data value on the bus (0 or 1).

Importance of Encoding

Automotive applications present unique constraints on bit encoding. The primary goal is low radiated emissions (radio-frequency interference—RFI). This is usually gained by a low transmission rate. Using an expensive medium such as a shielded twisted pair or fiber optics can also reduce emissions. Sometimes, it is feasible to add some complexity to the transceiver and do extensive wave shaping and/or filtering.

The simplest physical data link would be a single wire, randomly placed, and locally ground referenced. Also desirable is the fewest number of transitions possible per bit and per message, so as to get low radiated emissions.

Signal Encoding Methods

A few examples of commonly used encoding methods are shown here. In each case, an illustration is given of a typical logical implementation. Each graphic assumes an inactive or off-state (resting, or ground) of the bus. The "+" and "–" levels are driven (active) states. This driven state may be a voltage level, current level, or light on (fiber optic). If the bus is a two-wire implementation, the driven state may be a voltage differential.

When encoding is suitable for arbitration, the implementation may be made zero-dominant or one-dominant. All examples following assume one-dominant, except for the variable pulse-width (VPW) case.

Several desirable attributes are discussed for each encoding scheme. One of these is minimum feature size. The minimum feature size is the smallest pulse width that receivers must recognize. Especially if the protocol requires sampling to determine bus level, minimum feature size can have a large effect on noise rejection [1]. A small feature size is harder for receivers to recognize and is more susceptible to noise pulses.

Another feature compared among the methods is the ability to arbitrate. Another is fixed-length bits versus variable bit length. All encodings have fixed-length bits, except for VPW. The number of transitions per bit is also important. A large number of transitions is desirable from the viewpoint of resynchronization. However, RFI would dictate the fewest number of transitions possible.

Ratio (Pulse-Width Modulation)

Figure 5.1 shows an example of ratio, or pulse-width modulation (PWM), encoding.

PWM is a popular encoding method and is used in SAE J1850, for example. The "ratio" of time the bus is in the active state may be 1/3–2/3 (shown) or 1/4–3/4 of the bit time, or something else.

Note that there are two transitions per bit (one bit per bus period) and the minimum feature size is the smallest ratio (i.e., 1/3 of a bit in Fig. 5.1). These are less than ideal attributes. For example, VPW (discussed later) encoding carries two bits of information per clock period [2].

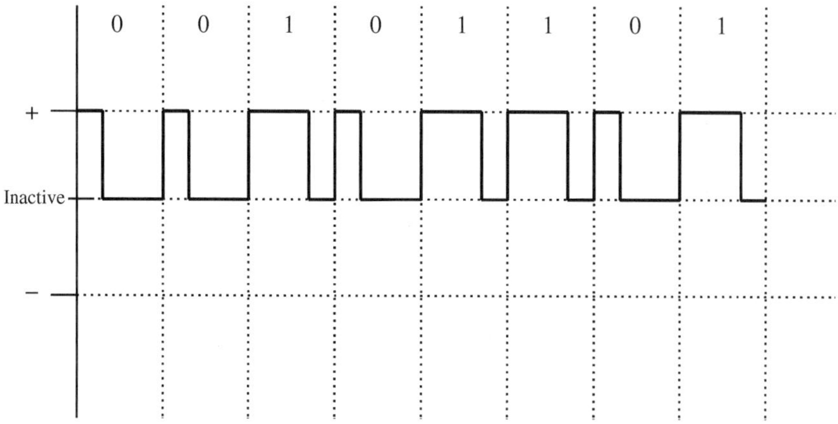

Fig. 5.1 Example of pulse-width modulation (PWM) encoding.

The following is a summary of the good/less-than-good points of PWM:

Ratio (Pulse-Width Modulation) Advantages:

- Fixed-length bits
- Defined sampling points
- Ability to arbitrate on contention buses
- Ability to resynchronize all receivers on the rising edge of each bit

Ratio (Pulse-Width Modulation) Disadvantages:

- Two edges per bit
- 1/3 bit time may be short time to decode in noisy conditions

Manchester-M (Biphase M)

There are several forms of Manchester encoding. One of the more popular forms is Manchester-M, or Biphase-M, illustrated in Fig. 5.2.

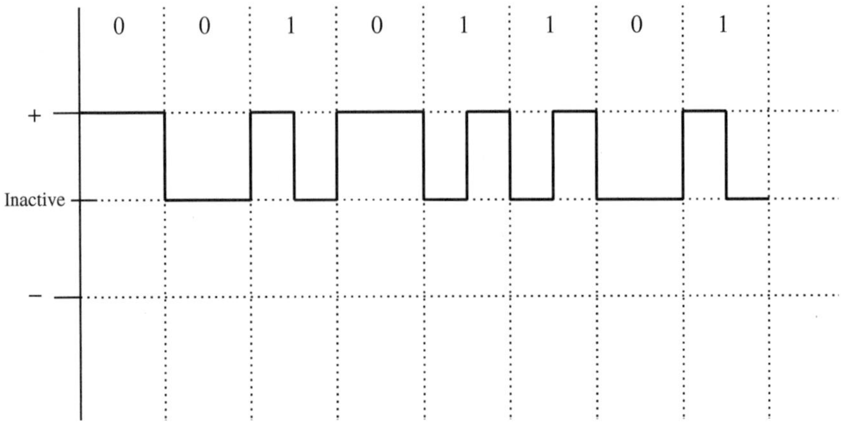

Fig. 5.2 Example of Manchester-M (Biphase-M) encoding.

In this example of Manchester-M, an encoded zero has no transition during a bit. A one is encoded with a transition midway through the bit time. This gives an average of 1.5 transitions per bit. This is better for RFI than the 2.0 average of PWM. The minimum feature size of 0.5 is also better than PWM.

Manchester-M Advantages:

- Fairly low number of transitions per bit
- Fixed transition (and sample) points
- Easy synchronization of bits

Manchester-M Disadvantages:

- Not suitable for arbitration

Manchester-L (Phase)

Manchester-L or phase encoding (Fig. 5.3) is similar to Manchester-M, except that a zero also has a transition during the bit. Ones have a falling transition,

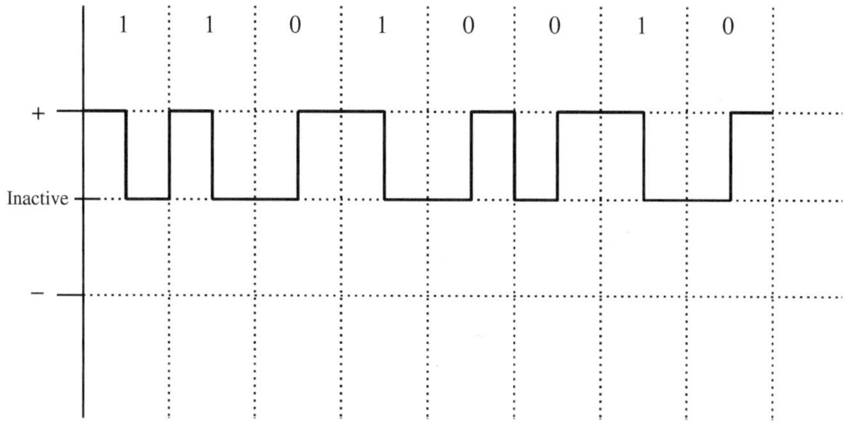

Fig. 5.3 Example of Manchester-L (phase) encoding.

and zeros have a rising transition. This gives a 1.5 transition per bit average. The minimum bit feature is still 0.5 of a bit period.

Manchester-L (Phase) Advantages:

- (Fairly) low number of transitions per bit
- Fixed transition (and sample) points
- Easy synchronization of bits
- Extra transitions needed between pairs of equal bits
- Suitable for arbitration

Manchester-L (Phase) Disadvantages:

- Small minimum feature size
- Too many transitions

Modified Frequency Modulation (MFM)

Modified frequency modulation (MFM) has not yet been used in automotive systems (Fig. 5.4). It is typically found in disk drive applications and is discussed here only as an example [3].

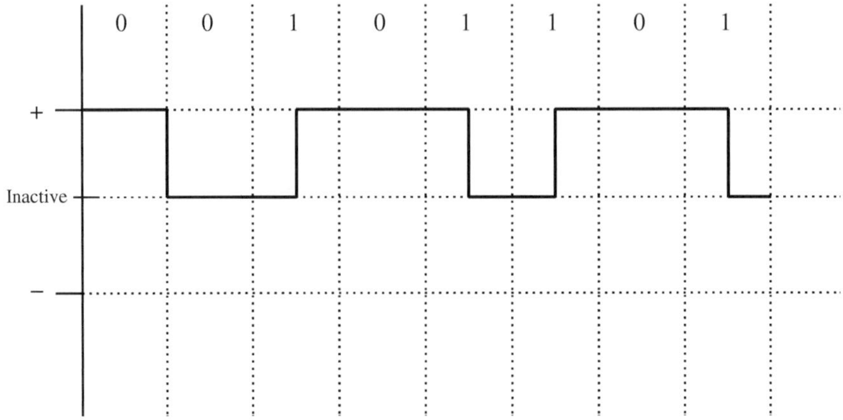

Fig. 5.4 Example of modified frequency modulation (MFM) encoding.

In MFM, one-bits are encoded with a transition midway through the bit. There is no transition within a zero, but there is a transition between zero bits. This results in a 0.75 transition per bit average and a minimum feature size of 0.5 bit.

MFM Advantages:

- Low number of transitions per bit
- Fixed transition (and sample) points

MFM Disadvantages:

- Small minimum feature size
- More difficult synchronization of bits
- Not suitable for arbitration

Return to Zero (RZ) Unipolar

Return to zero (RZ) is an old encoding that has been used for a long time in nonautomotive systems (Fig. 5.5).

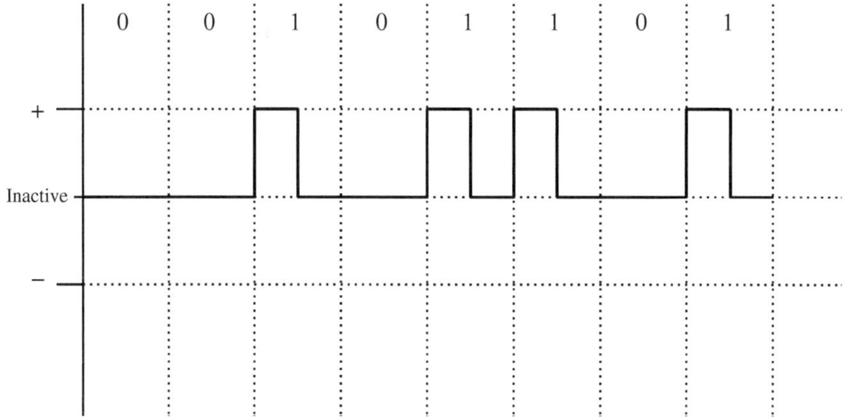

Fig. 5.5 Example of return to zero (RZ) unipolar encoding.

In RZ unipolar, zero bits have no transition either within the bit or between bits. One bits, however, have a transition back to the inactive state midway through the bit. There is an average of 1 transition per bit and a 0.5 bit period minimum feature size.

RZ Unipolar Advantages:

- Low number of transitions per bit
- Fixed transition (and sample) points
- Suitable for arbitration

RZ Unipolar Disadvantages:

- Small minimum feature size
- More difficult synchronization of bits

Return to Zero (RZ) Bipolar

Return to zero (RZ) bipolar is similar to RZ unipolar, except that two active states are used. This is so that both zeros and ones may have transitions back to the inactive state. The technique is best illustrated with a picture (Fig. 5.6).

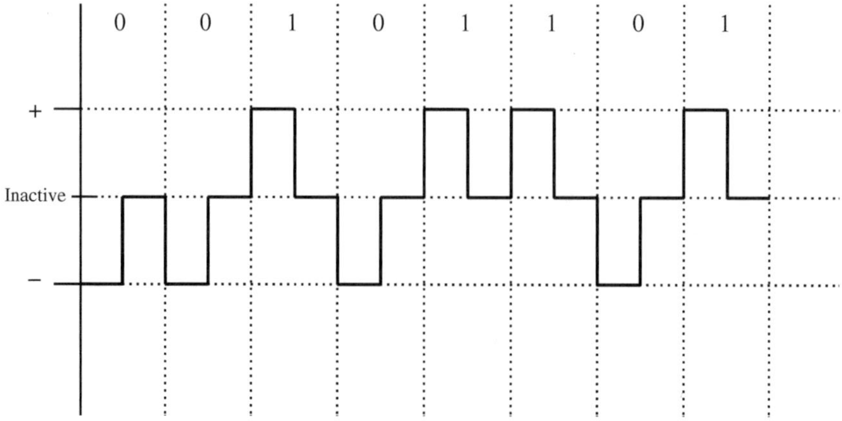

Fig. 5.6 Example of return to zero (RZ) bipolar encoding.

RZ bipolar has two transitions per bit and a minimum feature size of 0.5 bit.

RZ Bipolar Advantages:

- Fixed transition (and sample) points
- Suitable for arbitration

RZ Bipolar Disadvantages:

- Small minimum feature size
- Larger total voltage swing
- High number of transitions per bit

Nonreturn to Zero (NRZ) Unipolar

Nonreturn to zero (NRZ) unipolar is very important in automotive serial data communications (Fig. 5.7). This encoding method is commonly used with UART protocols.

NRZ unipolar has a certain feature that is a mixed blessing: the only transitions are between zeros and ones. If the data stream contains mostly, or all,

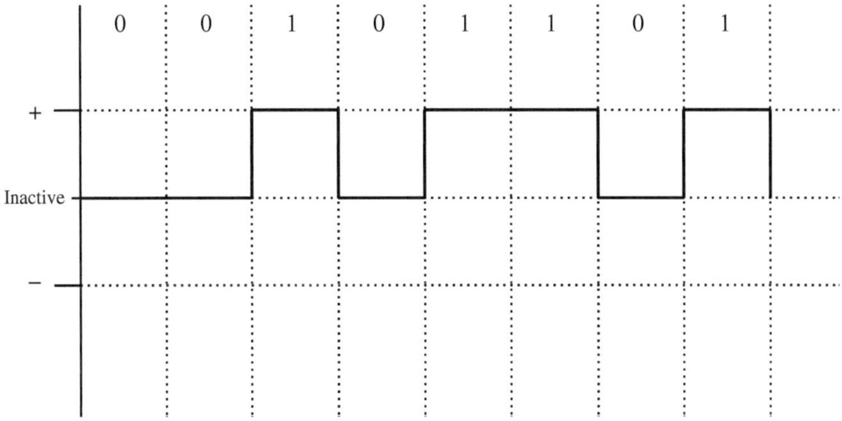

Fig. 5.7 Example of nonreturn to zero (NRZ) unipolar encoding.

one bit or the other, there will be few or no transitions. This causes problems with synchronization, but clever data encoding can work around this problem, as will be shown later.

The minimum feature size of 1 bit is very good.

NRZ Unipolar Advantages:

- Low number of transitions per bit
- Fixed transition (and sample) points
- Suitable for arbitration

NRZ Unipolar Disadvantages:

- Accumulation of clock errors among nodes
- Harder synchronization of bits

Nonreturn to Zero (NRZ) Bipolar

Nonreturn to zero (NRZ) bipolar is not really applicable but is shown here for completeness (Fig. 5.8).

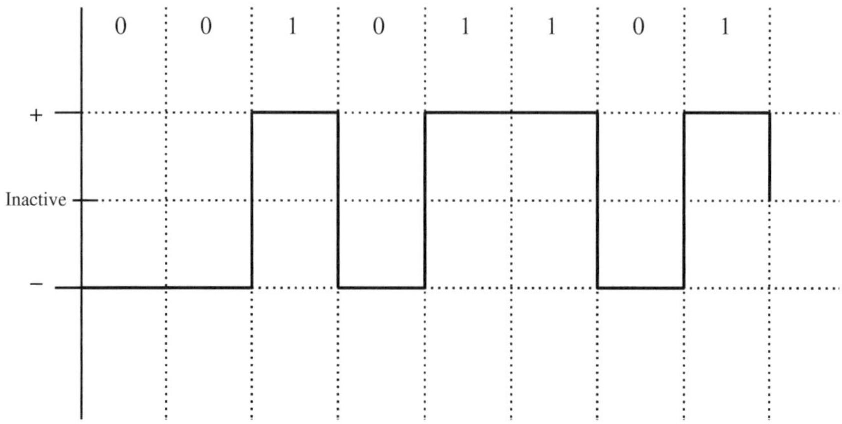

Fig. 5.8 Example of nonreturn to zero (NRZ) bipolar encoding.

NRZ bipolar has the same advantages and disadvantages as NRZ unipolar, except for larger voltage swings that make it more immune to noise.

Nonreturn to Zero (NRZ) with Bit Stuffing

Nonreturn to zero unipolar (hereafter referred to as NRZ) with bit stuffing is a very useful extension to NRZ.

Bit stuffing means that the transmitter inserts an extra, opposite bit into the data stream after a certain number of equal bits. This is to maintain enough transitions for the receivers to remain synchronized with the other nodes. A typical number of equal bits is five (NRZ-5), as used in the CAN and automobile bus (ABUS) protocols (Fig. 5.9).

Receivers know to throw away the next bit, if a specific number of equal bits occurs.

A case where stuff bits can occur every four data bits during NRZ-5 encoding is illustrated in Fig. 5.10. For this to happen, the data bits must come in groups of four equal bits after an initial stuffing sequence. The significance is that the stuff bits have become 20% of the data throughput. Although not likely to

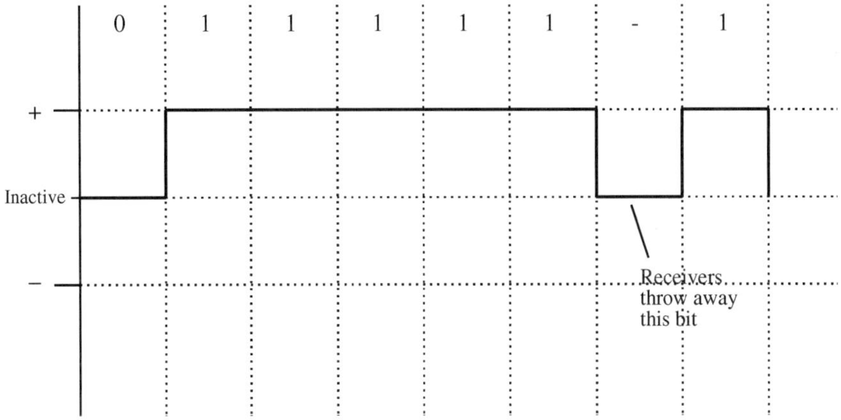

Fig. 5.9 Example of nonreturn to zero (NRZ) with bit stuffing encoding.

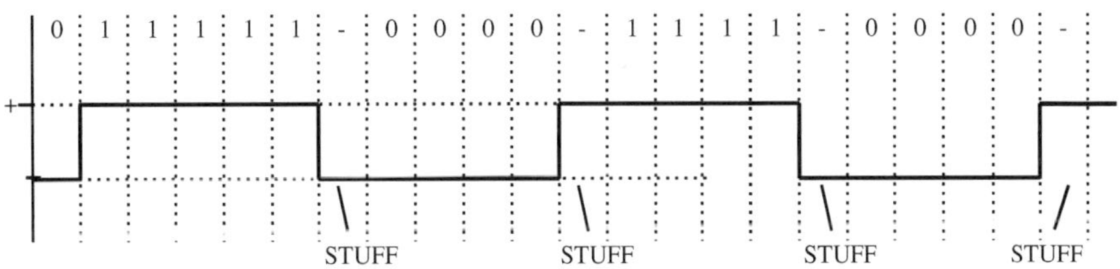

Fig. 5.10 Example of NRZ-5 special bit stuffing.

happen, this bit sequence may have to be considered when calculating the worst case message latency. Another consequence of this bit stream is that it is the longest time between resynchronizations in CAN. Thus, when using the typical method of resynchronizing only on recessive to dominant edges, there are ten bit times between such edges.

The probability that any bit in a data stream is a stuff bit is given by the following formula:

$$p_i = \begin{cases} \phi & \text{if } i < 5 \\ 1/32 & \text{if } i = 5 \\ 1/64 & \text{if } i > 5 \end{cases}$$

The number of stuff bits per message is

$$F(n) = (\Sigma)\ p_i$$

where

n = number of bits/message
i = ith bit position in the message

Table 5.1 calculates the output of the above formula for a one to twelve byte message.

It can be seen that, on average, there is approximately one stuff bit for every eight or nine data bytes. This does not have a large effect on latencies but does need to be taken into consideration when analyzing message delays.

TABLE 5.1
AVERAGE NUMBER OF STUFF BITS

Message Length (Bytes)	# Stuffed Bits (per message)
1	0.0625
2	0.1875
3	0.3125
4	0.4375
5	0.5625
6	0.6875
7	0.8125
8	0.9375
9	1.0625
10	1.1875
11	1.3125
12	1.4375

Manchester-E (Enhanced Manchester)

Manchester-E, or enhanced Manchester, is a relatively new development in signal encoding. It is most notably used in the vehicle area network (VAN) protocol. This encoding is unique in that it is a combination of two other encodings, Manchester-L and NRZ.

The data stream is broken up into four-bit chunks. The transmitter sends three bits as NRZ and each fourth bit as Manchester-L (Fig. 5.11). The Manchester bit is to guarantee an edge for synchronization. This is compared to bit stuffing as a method of aiding resynchronization. With Manchester-E, there is an edge every four bits. The minimum feature size is 0.5 bit, which is not so good, but there are also no extra bits inserted as with bit stuffing.

Manchester-E Advantages:

- Low number of transitions per bit
- Fixed transition (and sample) points
- Easy synchronization of bits
- Extra transition guaranteed by Manchester-L bit
- Suitable for arbitration

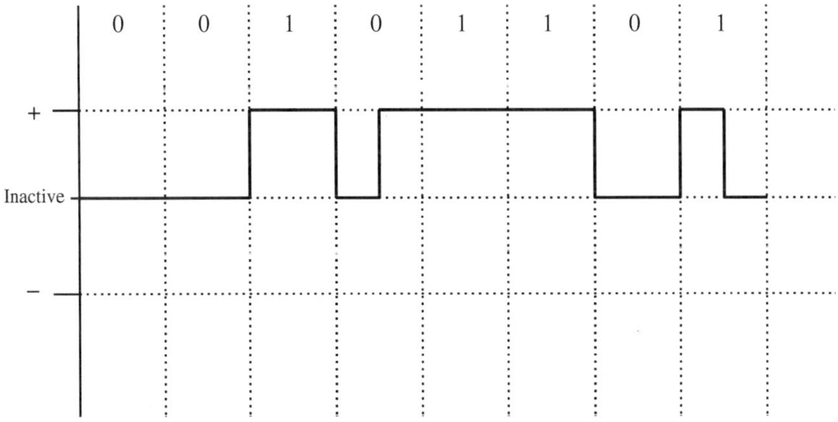

Fig. 5.11 Example of Manchester-E (enhanced Manchester) encoding.

Manchester-E Disadvantages:

- Small minimum feature size
- More complicated encoder and decoder

Variable Pulse-Width (VPW) Modulation

Variable pulse-width (VPW) encoding is a patented form of encoding invented by Fred Huntzicker of GM Research Labs. Sometimes, this encoding method is called "Huntzicker" encoding.

The key to understanding VPW (Fig. 5.12) is to note that zeros and ones are not directly determined from the voltage level. Whether a bit is a zero or one depends on the voltage level and the length of time at that voltage level. There are typically two voltage levels (some positive level and ground) and two time lengths (short and long). For example, in VPW as used in J1850, a zero bit is a short low pulse or a long high pulse. A one bit is a long low pulse or a short high pulse. There are two bits per bus period, and the bus switches voltage level between each bit.

The minimum feature size in VPW is the width of the short pulse. There is one transition per bit, which is very desirable, but the data rate is not constant. For example, in J1850, the first bit is in the low state, and a stream of $55

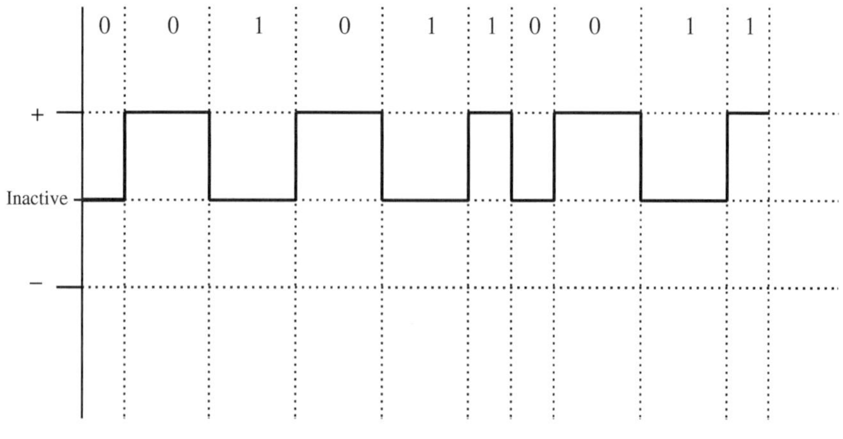

Fig. 5.12 Example of variable pulse-width (VPW) encoding.

bytes would go by at 15.6 kb/s, whereas a stream of $AA bytes would be going at 7.8 kb/s. The *average* data rate is 10.4 kb/s, but the instantaneous data rate depends on the data content. This wreaks a little havoc in calculating message latencies.

VPW Advantages:

- Low number of transitions per bit
- Fixed transition (and sample) points
- Well suited for arbitration
- Can compensate for clock mismatch and ground offsets

VPW Disadvantages:

- Not intuitively obvious
- Symbol timing very important for receivers
- Data rate is variable

An example of contention resolution for VPW is shown in Fig. 5.13.

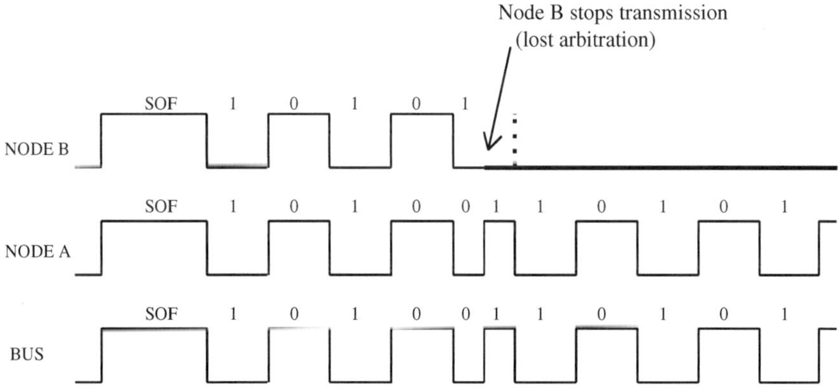

Fig. 5.13 Example of VPW arbitration.

An example of how edge mismatch for VPW is compensated for is shown in Fig. 5.14. The final bus waveform is a wired-OR of all transmitters active at the same time. Arbitration can be described as a wire-AND of all current transmitters.

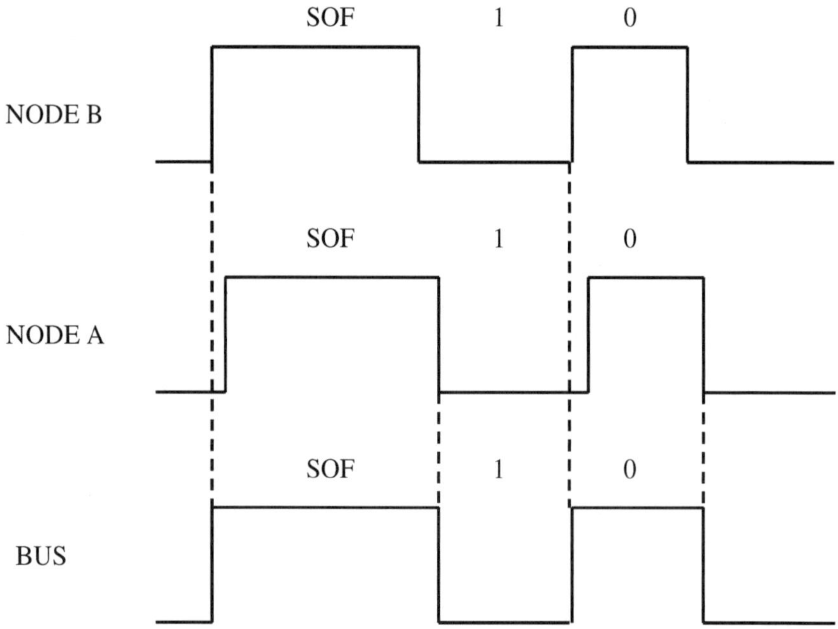

Fig. 5.14 Example of VPW mismatch.

Figure 5.15 gives a pictorial summary of the symbol encoding methods already discussed. VPW is not included because it does not have fixed-width bits and will not line up with the bit boundaries of the other methods.

References

[1] Dais, S., and Chapman, M. "Impact of Bit Representation on Transport Capacity and Clock Accuracy in Serial Data Streams." SAE Paper No. 890532, Society of Automotive Engineers, Warrendale, PA, 1989.

[2] Hu, C., Wills, A., Pogue, R., and Carman, D. "Serial Bus Thruput Doubling by Using Variable Pulse Width Modulation." SAE Paper No. 910714, Society of Automotive Engineers, Warrendale, PA, 1991.

[3] Miesterfeld, F., and Halter, R. "Survey of Encoding Techniques for Vehicle Multiplexing." SAE Paper No. 910715, Society of Automotive Engineers, Warrendale, PA, 1991.

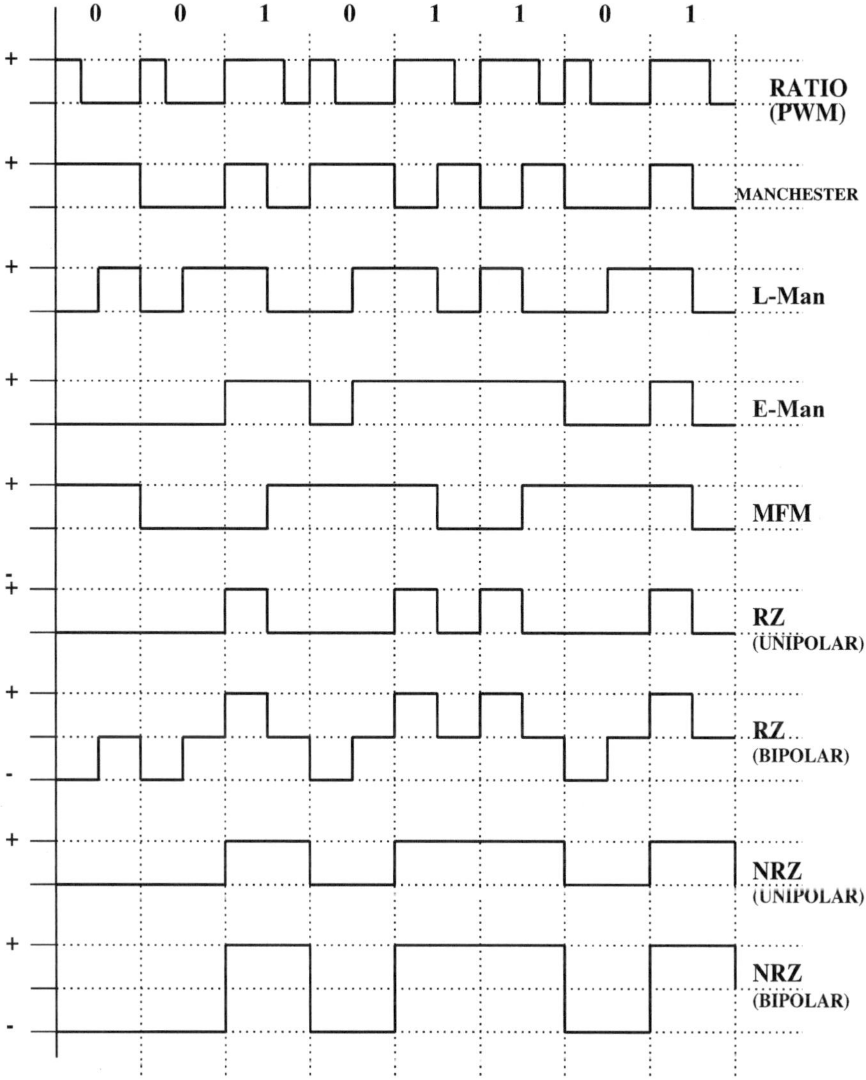

Fig. 5.15 Summary of encoding.

Chapter 6

Error Handling

Introduction

Error handling is important in a network's handling of inevitable faults. Several general topics are discussed, followed by a description of the five major categories of error handling:

- Error reporting
- Error recovery
- Error detection/protection
- Error correction
- Error confinement

From Where Do Errors Come?

There are numerous sources of errors, but they can all be categorized as originating either within the vehicle or outside the vehicle. Two key sources of errors inside the vehicle are noise from the ignition system and any radio transmitters (e.g., mobile transmitter, cell phone). The two primary sources of interference outside the vehicle are radio transmission towers and adjacent vehicles.

Other sources of errors inside the vehicle are from intermittent wiring connections as well as broken or shorted connections.

Error Types

Errors are categorized as local or global. A global error is one that is detected by all nodes. A local error is a glitch or disturbance that is detected by only some of the nodes. This can happen because of bit sampling variations due to propagative delays or different receiver switching thresholds. Noise is weaker at the node farthest from the source and may not affect all nodes.

Local errors are much more dangerous than global errors. If all nodes see the error, then there is usually no problem. A local error implies that one or more nodes have accepted potentially faulty or corrupted data from the bus.

Noise is categorized by whether it is "micronoise" or "macronoise." Micronoise affects only a small portion of the bit stream, probably less than one bit (Fig. 6.1). This type of noise can cause a local error, depending on where the sample point is within the bit. Not all nodes will sample the bit at exactly the same point because of variations in the receiver threshold and propagative delays.

Macronoise, on the other hand, wipes out large portions of the data stream (Fig. 6.2). This type of disturbance is usually much easier to detect and almost always causes a global error.

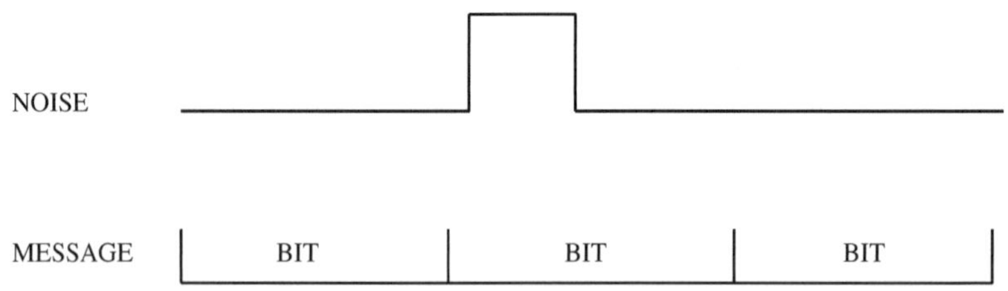

Fig. 6.1 Micronoise.

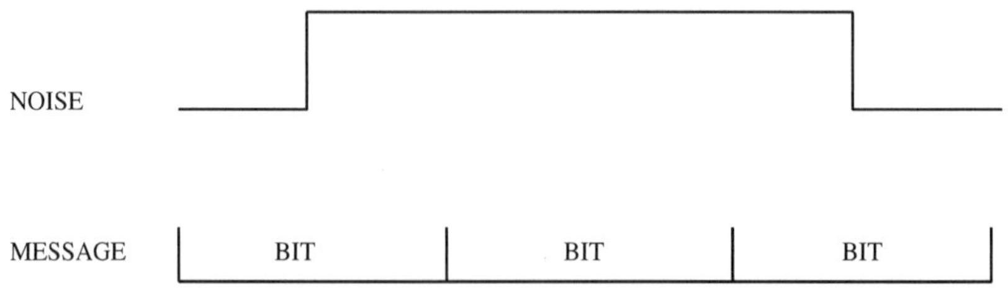

Fig. 6.2 Macronoise.

Failure Rates

Failure rates are hard to predict. This is a whole field of study in itself. The following example illustrates that a serial data network is more reliable than a bundle of discrete wires carrying the information.

The first formula is for a discrete bundle of k wires:

$$p_s = k(p_{TR} + p_{RC} + 2p_C + p_L)$$

where

p_s = probability that one of k separate wires fails
p_{TR} = probability that transmitter circuit fails
p_{RC} = probability that receiver circuit fails
p_C = probability that connector fails
p_L = probability that wire itself fails

The next formula is for a serial data bus:

$$P_B = n(p_{TR} + p_{RC} + mp_C) + mp_L$$

where

p_B = probability that serial bus connecting n nodes by m wires fails
p_{TR} = probability that transmitter circuit fails
p_{RC} = probability that receiver circuit fails
p_C = probability that connector fails
p_L = probability that wire itself fails

If we assume that k is much larger than n, and m is equal to two (two-wire bus), then it can be shown that the serial data network is more reliable than a bundle of separate wires. It must be mentioned, however, that single-point failures can be more catastrophic in the serial data case.

Error Detection and Protection

Detection of errors is the most important aspect of dealing with multiplex faults. Although error rates are hard to predict, the goal is to detect all hardware faults. After that, one has to deal with error recovery, fault confinement, and so forth, but preventing undetected errors is paramount.

Many things can be done to deal with errors when they are detected. Some examples are resending the data, updating error counters, and alerting the vehicle operator.

There are two classes of errors that can be analyzed—detectable and undetectable. Detectable errors are the best kind. They can be dealt with and ultimately should have no effect on the overall operation of the network. The frequency of occurrence of these errors is very hard to predict. An acceptable rule of thumb is that between 1 in 10^3 and 1 in 10^5 messages will be disturbed. This is an average during the vehicle's lifetime and includes noisy and nominal operating cycles. Most disturbances will be caused by outside forces such as radio transmission towers and electrical plants, as pointed out previously.

Undetectable errors are a class of faults that are extremely difficult to analyze and predict because they are simply that—undetectable. You can theorize how an undetectable error can occur, but you cannot trap or trigger on it. Several undetected error studies have been done in the past for J1850 and CAN. Several years ago, Ford commissioned the University of Michigan to do a study on two-wire J1850, and GM had Hughes Aircraft Company do another for one-wire J1850. These proprietary reports reached conclusions as to the residual error probability for undetected errors but were much debated.

Analyzing undetected errors requires many assumptions about error rates, error probabilities, environment, hardware technology used, and so forth. These assumptions ripple through the analysis and have a large effect on the final outcome. There are also usually several types of errors that may be undetectable. For example, for CAN, it is assumed that there are three types of undetectable errors that can happen [1]. The first is from normal bit errors due to the finite coverage of the cyclic redundancy check (CRC). The second is from encoding errors where information bits are turned into stuffed bits, and vice versa. The third type is caused by message-modifying errors that occur when the data length code is corrupted or an end-of-frame symbol is accidentally created. This last type is very important in most serial data protocols. The Class 2 protocol (discussed elsewhere in this book) has a unique function for dealing with message length errors.

A generally acceptable rate of occurrence for undetected errors is less than one per vehicle lifetime. Today, this lifetime is about 4000 hours or 100,000 miles. The following example gives an idea of the undetected error rate necessary.

Assuming

r_e = 10^{-3} message error rate
r_t = 150 kb/s bit rate
l_t = 50% bus utilization
l_m = 80 b/message average message length

then

$r_m = r_t \times l_t / l_m = 940$ messages/s
=> 10^{10} messages per vehicle lifetime (3000 h)
$n_e = r_e \times r_m \times 3000\text{ h} = 10^7$ total number of errors

Requiring no more than 1 undetected error per 3000 h,

$$n_{nd} \leq 1$$

The "residual error probability" works out to

$$p_{nd} = n_{nd}/n_e \leq 10^{-7}$$

The residual error probability tells "how good" the protocol has to be to avoid undetected errors. In this case, we desire that the protocol have no more than 10^{-7} (e.g., 10^{-8}) chance of an undetected error.

According to a 1990 paper [1], the calculated residual undetected error probability of the CAN protocol is on the order of 10^{-14}. This makes CAN a very robust system.

Several schemes are used in vehicle multiplex networks for error detection. Most protocols use more than one.

Although parity is common in the computer industry, it is not typically used in automotive serial protocols. A checksum is used by GM's Class 2 protocol (in addition to a CRC) during block mode.

The CRC is a powerful error detection mechanism that is well suited to automotive applications. It is much more efficient than parity, especially for long messages. CRCs are typically 8 or 16 bits long but may be any length. A CRC is basically a code that the transmitter appends to the end of every message. When the receiver processes the incoming message, including the CRC, a certain result is expected through its check circuit. If the result is different from what it should be, the message is marked as an error or is discarded.

The CRC "code" is the remainder of a modulo-2 arithmetic operation between the message frame and the CRC polynomial. It begins with the remainder register set to all ones. Then the outgoing message frame is divided, bit-by-bit, by the CRC polynomial. The remainder is the CRC.

Transmitting nodes calculate the CRC on the outgoing frame. Then, they invert the remainder and append the result to the frame. Receiving nodes do the same division operation on the incoming bit stream including the inverted CRC. This result will always be a fixed value if the frame was not corrupted.

CRC polynomials are typically picked so that the CRC check can detect the following errors:

- All burst errors of length n or less
- All burst errors of length n + 1 except one (the bit pattern of the CRC polynomial itself)
- All one- or two-bit errors
- All odd numbers of bits in error
- More than 99% of all other errors

Basically, the CRC is the extension of a frame that makes the frame evenly divisible by the polynomial. Most disruptions of the frame contents ruin this relationship and make the error obvious.

CRC calculations are much easier to do with hardware than with software, mainly because of the difficulty of modulo-2 arithmetic operations. CRCs are very easy to accomplish with hardware. Figure 6.3 shows a representation of a CRC generator/checker circuit often used with the J1850 protocol. Below the shift register implementation is an example of a CRC calculation for the hexadecimal data stream "F2h 01h 83h." The data stream is clocked through the CRC latches, and the result is "C8h," which is the CRC. The calculation continues as if the data stream were being received at some other node. This receiver processes the data stream plus the inverted CRC (37h) and comes up with the remainder "C4h." This is always the expected result. If any receiver ever gets a remainder different from C4h, there is a problem with the frame.

An interesting thing to note about CRC circuitry is that in half-duplex systems (such as CAN), the hardware is usually shared between the transmitting and receiving portions of a node. In other words, there is not a separate CRC generator and CRC checker. The result is that a transmitting node actually does its calculations on only the outgoing bit stream. The CRC is stuffed into the shift register after the last data byte, so that the calculation comes out OK. This is considered acceptable because not much should happen between the transmitting node and the bus.

Several other schemes are used for error protection in vehicle multiplex networks. Like error detection, most protocols use more than one method.

Analog filtering is common in transceiver circuits to prevent noise from corrupting the incoming bit stream. Digital filtering is common in hardware for the J1850 VPW (GM Class 2) protocol because encoding it is based on symbol timing anyway. Filter widths of 8 or 16 μs are usually used to ignore pulses shorter than that. An issue with this kind of filtering is the fixed delay it adds to the outgoing data stream, which may be an issue for arbitrated protocols. Majority sampling can be categorized as a type of digital filter. Using this

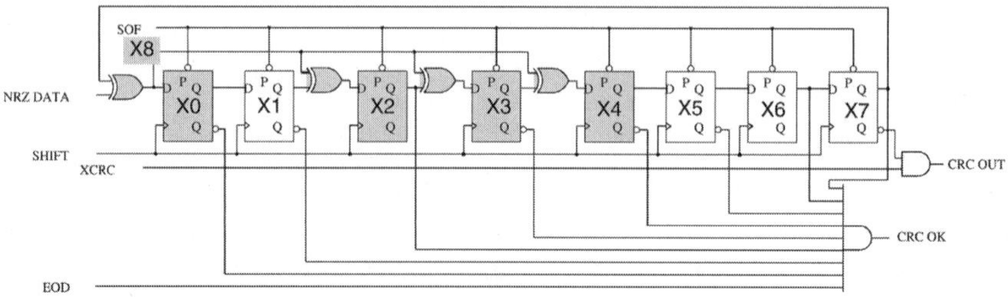

EXAMPLE MESSAGE = F2 01 83 (CRC=37) Generator = X8+ X4+ X3+ X2+ X0

SHIFT		NRZ DATA	X7	X8	X0	X1	X2	X3	X4	X5	X6	X7	
Start			1	-	1	1	1	1	1	1	1	1	
1		1	1	0	0	1	1	1	1	1	1	1	
2		1	1	0	0	0	1	1	1	1	1	1	
3		1	1	0	0	0	0	1	1	1	1	1	
4	F2H	1	1	0	0	0	0	0	1	1	1	1	
5		0	1	1	1	0	1	1	1	1	1	1	
6		0	1	1	1	1	1	0	0	1	1	1	
7		1	1	0	0	1	1	1	0	0	1	1	
8		0	1	1	1	0	0	0	0	0	0	1	
9		0	0	1	1	1	1	1	1	0	0	0	
10		0	0	0	0	1	1	1	1	1	0	0	
11		0	0	0	0	0	1	1	1	1	1	0	
12	01H	0	1	0	0	0	0	1	1	1	1	1	
13		0	1	1	1	0	1	1	0	1	1	1	
14		0	1	1	1	1	1	0	0	0	1	1	
15		0	1	1	1	1	0	0	1	0	0	1	
16		1	0	0	0	1	1	0	0	1	0	0	
17		1	0	1	1	0	0	0	1	0	1	0	
18		0	1	0	0	1	0	0	0	1	0	1	
19		0	0	1	1	0	0	1	1	0	1	0	
20		0	1	0	0	1	0	0	1	1	0	1	
21	83H	0	0	1	1	0	0	1	1	1	1	0	
22		0	1	0	0	1	0	0	1	1	1	1	
23		1	1	0	0	0	1	0	0	1	1	1	
24		1	1	0	0	0	0	1	0	0	1	1	CRC
25		0	1	1	1	0	1	1	0	0	0	1	
26		0	0	1	1	1	1	0	0	0	0	0	
27	CRC	1	0	1	1	1	0	0	1	0	0	0	
28	37H	1	0	1	1	1	0	1	1	1	0	0	
29		0	0	0	0	1	1	0	1	1	1	0	
30		1	1	1	1	0	0	0	1	1	1	1	
31		1	1	0	0	1	0	0	0	1	1	1	
32		1	1	0	0	0	1	0	0	0	1	1	C4H

Fig. 6.3 J1850 CRC generator/checker circuit.

scheme, more than one voltage sample is taken during the bit time—quite a few in some protocols. A percentage threshold must be met for the bit to be decoded as a one or zero, say, two out of three samples must match.

Bit stuffing is another error protection mechanism that has been discussed in Chapter 5.

A simple method for protecting against errors from noise is to increase the amplitude of the signal. This makes the signal easier for receivers to decode in the presence of noise. It also adds, however, to the radiated noise component.

In general, there are several types of errors that have to be detected. Most of the time, they are set forth in the protocol specification. The main ones are bit errors, byte errors, and framing errors. To detect bit errors, the receive logic checks to determine that each bit was a valid bit and also that it had the correct polarity (zero or one), if possible. Obviously, the bit value detected may not match the value sent during the arbitration phase (if a contention bus). A valid bit will have the proper voltage level and/or correct length.

A byte error occurs if the message contained any bytes that were not 8-bit chunks. This check is primarily in the data field of the message.

The receive logic must check for correct fixed fields of the message. Usually this involves verifying reserved bits in the message, any start or end bits, and so forth.

Error Reporting

Various error-reporting methods are employed in serial data protocols. One of them is the concept of acknowledgment. There are two kinds of acknowledgment, or "ACK." The first is positive ACK. This means that receivers respond actively in some fashion to indicate correct receipt of the message, even if they were not the intended targets for the message. No ACK indicates to the transmitting node that nobody received the message correctly.

Negative ACK is usually the "no news is good news" method. If nobody complains about the message, the transmitting node assumes that there was no error. This is simpler than positive ACK but is not suited for Class C networks.

In-frame response (IFR) is a scheme employed in several automotive serial data protocols such as J1850 and VAN. An IFR is an extension of a message or is embedded in a message by some node other than the original transmitter. Obviously, the IFR transmitter has to be quick to get its IFR into the data stream at the right time. This brings up the major argument against the use of IFR—the receiver has not had time to process the original message before responding. Typically, there is time to only check the message for errors and append a physical source identifier as the IFR. Ford and DaimlerChrysler employ this strategy for some diagnostic operations with J1850.

CAN uses an error-reporting method that involves error frames. An error frame is a special message sent out by any node(s) that detects an error. The basic purpose of the error frame is to prevent a local error. Error frames violate the bit-stuffing rule of CAN and thus force all nodes to detect an error (because of the error frame!). All other nodes then send an error frame, causing a "superposition" of error frames.

Error Recovery

Error recovery is important in networks and is often set forth in the protocol specification. This includes the way error recovery is to be done, as well as the maximum time to recover from an error and resume normal bus communication.

Short recovery times are desirable, especially in contention buses, so that high-priority messages get out quickly. Contention protocols usually suggest automatic retransmission of a message in case of an error during the first try. CAN is such a protocol. This provides for a very short recovery time compared to a token–slot or master–slave type of system. For a token-passing network, for example, the entire bus may have to go through initialization again, and the token(s) must be reproduced by some node.

Error Correction

Error correction is really beyond the scope of automotive multiplexing. It is, however, getting some attention for some future Class D applications such as video/audio transfer. An application like this requires accurate, high-bandwidth, real-time data. The targeted data rate for such systems is in the range of 100 Mb/s and will carry packeted or isochronous data.

Error correction is needed primarily in high-noise environments. In these systems, any retransmitted information is as likely to be corrupted as the original data. It is also necessary, obviously, in real-time networks. Another case is long-latency networks where data errors are very likely.

Reed–Solomon is one method of error correction. A circuit to implement a Reed–Solomon code takes a varying amount of logic, depending on the level of correction provided. In some cases, this can be as many as 50,000 active elements.

For automotive use, one assumes that errors are going to happen, and the goal is to be as fault-tolerant as possible. One also assumes that noise in the automotive environment is generally random and nonconstant, so error correction is not necessary.

Fault Confinement

Faults must be contained. Intermittent errors are usually not a problem because of retry schemes and other mechanisms. However, persistent problems should be dealt with. The CAN protocol does this by requiring that defective nodes "get off the bus" (bus-off state).

CAN isolates bad nodes by having the node decide that the errors it is experiencing are often and permanent. At this point, the node is not allowed to transmit or receive messages from the network. It does, however, monitor the bus for certain long periods of bus inactivity. After enough bus idle time, the node may rejoin bus communications. A hardware reset is really the only other way for a node to get back on the bus after going bus-off.

How CAN determines which node is responsible for an error is interesting. Basically, it involves "shooting the messenger" by penalizing the node that reports the error first. This node increases its receive error counter by a greater amount than the other nodes.

In general, all nodes on any network must use the same error handling and fault confinement strategy. Sometimes this is difficult if some nodes cost very little, have low functionality, and cannot afford the 8K to 12K of Read Only Memory (ROM) space to implement the full protocol handling requirements.

References

[1] Unruh, J., Mathony, H., and Kaiser, K. "Error Detection Analysis of Automotive Communication Protocols." SAE Paper No. 900699, Society of Automotive Engineers, Warrendale, PA, 1990.

Chapter 7

Media Characteristics

Introduction

This chapter discusses the commonly used wiring and fiber-optics technologies used in automotive serial data networks.

Twisted Pair Cables

Twisted pair is probably the most common medium used for automotive multiplex applications [1]. It is economical and produced in vast quantities. Although the bandwidth of a twisted pair is of the order of 5 MHz, concerns for electromagnetic compatibility (EMC) limit the useful maximum frequency to about 250 kHz.

The twisted pair is most commonly unshielded (UTP) but is also available shielded (STP). STP can be used in applications at frequencies higher than 250 kHz to reduce radiated emissions and susceptibility to radiation. A big problem with STP is maintaining the shield integrity for the lifetime of the vehicle. Even small openings in the shield, such as a connector after a repair, severely degrade the radio-frequency (RF) emissions performance of the network.

The wire pair does not have to be twisted; good performance can be had from "zip" cord. This is approximately the same thing as lamp, or speaker, cable. It is important that the two conductors remain in constant proximity to each other across their entire length to minimize the "loop area" formed by the two wires.

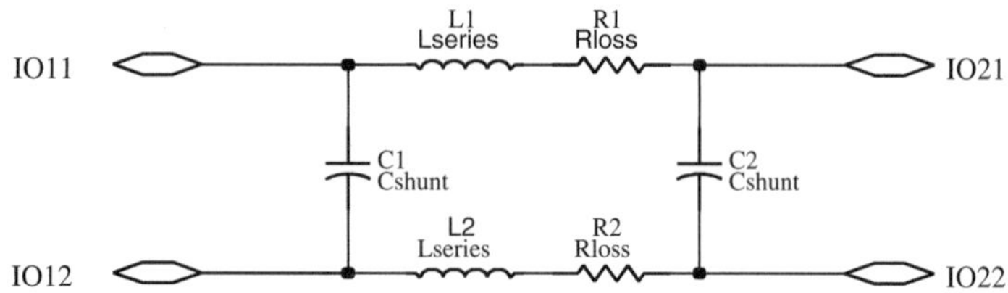

Fig. 7.1 Model of a basic unshielded twisted pair (UTP).

Figure 7.1 shows a typical wire model used for unshielded twisted pair. The following is an example of common values used for the components of the model:

Resistance: 70.0 m Ω per meter, series
Capacitance: 22.6 pF per meter, line-to-line
Inductance: 0.325 μH per meter, series

Coaxial Cables

Coax has a very high bandwidth. The bandwidth, and thus channel capacity, is on the order of 400 MHz. This medium has low radiated emissions because of its shield and is the typical cabling for the TV industry (i.e.,75 Ω impedance). Unfortunately, coax is relatively expensive and is considered mature, so little additional cost reduction is expected. There is no usage, currently, in automotive multiplex applications.

Fiber Optics

This section presents some information on fiber optics, a whole field in itself. References [2–17] provide more complete details.

The reasons for the market pull for optical fiber for interconnects in automobiles range from the high-tech image to reduced RFI/EMI and reduced mass. There is also the perceived need for much higher bandwidth capability in the future. The main impediment is cost. To replace an existing technology (copper wire in this case), the new technology must either cost less or provide some very visible benefit.

For the distances and data rates needed on a vehicle, plastic optical fiber (POF) and light-emitting diodes (LEDs), rather than glass fiber and lasers, are more than adequate. POF is very cost competitive with twisted and/or shielded wire suitable for high-speed signals. POF's large aperture allows for greater tolerance in the connectors, making them competitive with electrical connectors. The electronics is determined more by the data needs of the nodes than by the transmission medium. This leaves the LED and photo detector as the primary cost adders.

Data rates are limited by the LEDs to something over 100 Mbps. Once the decision is made to use fiber, this speed has been paid for. The best way to justify the added cost is to need the speed. Video applications such as rear passenger entertainment may be among the first to need the speed (about 10 Mbps for DVD compressed video), but audio applications can take advantage of it. Many applications can take advantage of high bandwidth (and an unlimited amount can be creatively wasted) until it becomes a real requirement at some point. This is when we will see widespread application of optical fiber in cars.

The first applications are in the passenger compartment, where the need for very high operating temperature is deferred. Normally, the temperature range used in product qualification testing of passenger compartment products is –40 to +85°C. LED brightness is a very important parameter. Better LED structures and/or more efficient optical coupling to the fiber can achieve it. It can be taken advantage of to allow longer fiber, in-line connectors, and passive splitters, and to increase noise margin. It can also be traded for lower power and possibly improved reliability. For the detector, the corresponding parameter is high sensitivity with high signal-to-noise for all the same reasons. Both need to be small to minimize connector size and improve packaging flexibility. From the packaging perspective, it is more important for the devices to be intimately coupled to the fiber than to the circuit board. Electrical isolation between the LED and detector is also an issue.

The MOST (media-oriented systems transport) bus is in production today. MOST is based on a ring topology. The advantages of the ring architecture are that it uses the minimum possible number of electro-optical devices (one receiver and one transmitter per node) and is the lowest cost implementation. The disadvantages of the ring architecture are fault tolerance (the network fails if any one node fails) and that the entire network must be up for any two nodes to communicate.

The Automotive Multimedia Interface-Collaboration (AMI-C) developed the IDB1394 automotive specification. IDB1394 is based on a tree topology, and it is compatible with fiber for the physical layer. The major disadvantage of tree architecture is that it requires more than the minimum number of electro-optical devices (one pair per tree branch), which increases the cost.

A plastic fiber optic network does not propagate signals faster than an electrical wire, contrary to what you might first expect. Because of the attenuative properties of the fiber on light and the attenuation of wire on current, both media transmit an electromagnetic signal at approximately 5.5 meters per nanosecond.

Fiber optics allows data rates of 100 Mb/s and more. These higher bit rates are possible mainly because the slew rate of the signal does not have to be limited.

The fiber itself is lightweight and generally not sensitive to chemicals or corrosion. The typical fiber that will be used is 1-mm diameter plastic with a plastic jacket. Although priced at about the same as coax, there is more room for the cost of the fiber to come down as the number of applications grows.

The largest cost components of a fiber optics network are the opto-electrical interface and the connector. The LED or laser diode for the transmitter is expensive—especially underhood temperature parts. This is also true for the PIN (p-doped intrinsic and n-doped) diode receiver. The basic definition of a PIN diode is as follows:

- A diode with a large intrinsic region sandwiched between p- and n-doped semiconducting regions. Photons absorbed in this region create electron–hole pairs that are then separated by an electric field, thus generating an electric current in a load circuit.

Because the fiber does not have the same "form factor" as a wire, there are connector difficulties when adding fiber optics to an ECU. The existing connector may be modified to accept the fiber(s) by using two wire cavities for each fiber, for example. Alternatively, and more commonly now, fibers get their own connector mainly so that the LED and PIN diode may be mounted within the connector and in-line with transmit and receive fibers. Sealed or "weatherpack" fiber-optic connectors are available now for underhood and undercar applications.

Although the fiber neither generates nor receives electromagnetic interference, the optical–electrical interface is still an area of potential noise.

Another issue in using optics is that the signal is unidirectional (half duplex) versus electrical signals, which are bidirectional (full duplex). Because of this, two fiber-optic cables are used, one to transmit and one to receive. A single cable would require a complicated reflector for both the LED and PIN diode to be "in-line" with the fiber (see Fig. 7.2). Another alternative would be to use wavelength division multiplexing and send multiple data streams at different light frequencies—an expensive proposition.

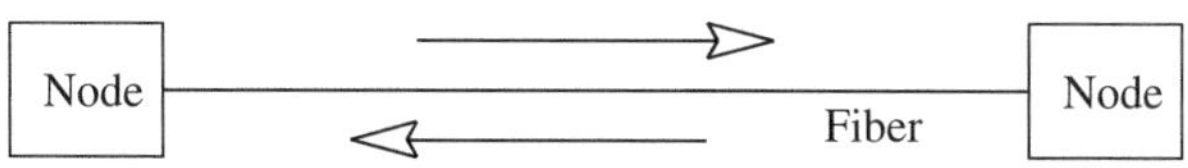

Fig. 7.2 A two-node, single-fiber network.

A laser diode is the transmitter of choice for fiber-optic systems. Red, the usual color, has a wavelength of about 660 nm. This is a rather large wavelength but is satisfactory for the slow signal rates and short distances in automotive applications. Green or blue laser diodes would allow faster data rates and better efficiency, but this is not a big issue. For vehicle networks, propagative and other electronic delays are dominant.

As mentioned previously, the typical fiber-optic cable employed is 1 mm in diameter for the best cost and weight. Plastic has much higher losses (about 0.2 dB per meter) than glass but is "good enough" for automotive uses.

When designing a fiber-optic system, the network developer uses several "flux budgets." The "optical power budget" is the amount of usable light in the passive network. It is basically the difference between the light strength from the weakest transmitter and the light needed by the least sensitive receiver. The "loss budget" is the maximum attenuation through the network from the transmitter and receiver that are farthest apart. A rule of thumb is to maintain at least 3 dB between the optical power budget and loss budget.

A two-node fiber-optic network is generally no problem to implement (see Fig. 7.3).

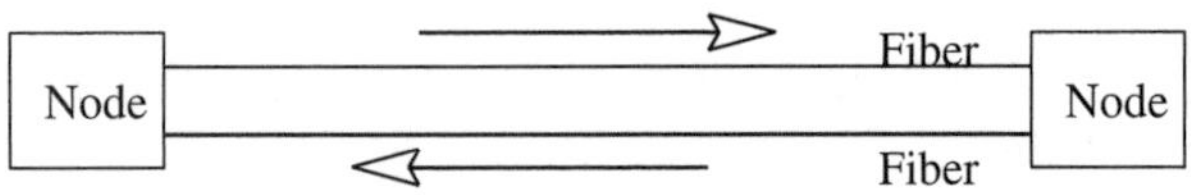

Fig. 7.3 A two-node, dual-fiber network.

When there are more than two nodes, the picture gets complicated. There are several ways to connect multiple nodes on a fiber-optic bus, but the one probably most suited for automotive is a star configuration. At the center of the star is a device called a fiber-optic star (Fig. 7.4).

The star connector passes the light signal from any node to the receive line of every other node, including the node transmitting the signal. This connector typically has two fibers attached to it for each node of the network. Years ago, the number of nodes that could be supported was fixed, and there was no way to add more nodes. For example, some early star connectors were made with seven transmit/receive pairs, or legs. If the application had four nodes, then three legs were wasted. If the application had eight nodes, then another connector had to be purchased. Today, "pluggable" star connectors are available that make it easier to add/delete nodes.

There are two main types of star connectors, or "stars." A passive star uses no amplification and can be either reflective or transmissive. A reflective passive star, you guessed it, reflects the incoming light in one fiber out to all outgoing fibers. A transmissive passive star splits the incoming light and sends it out on all outgoing fibers.

An active star would amplify the incoming light for retransmission out the other legs. This approach is considered too costly for automotive applications.

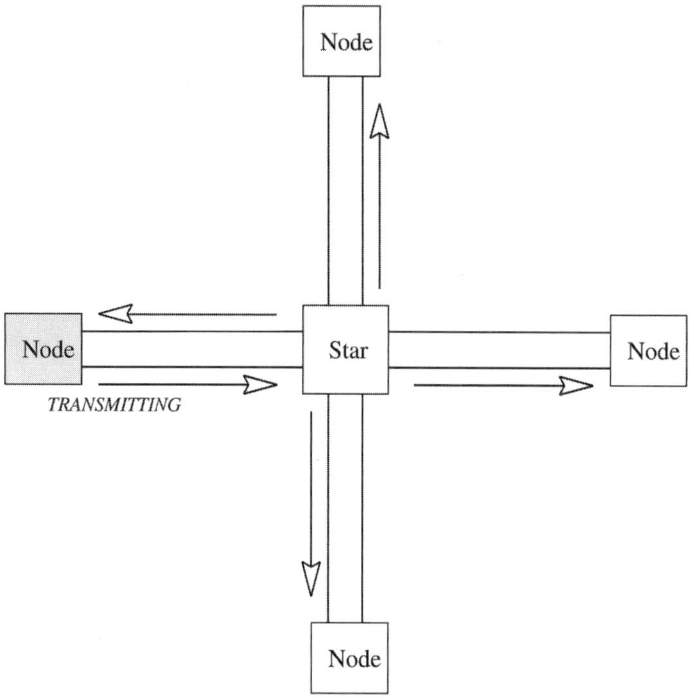

Fig. 7.4 Fiber-optic star configuration.

A couple of variations on the transmitter and receiver circuits are used with fiber optics. A transmitter configuration may be either shunt or nonshunt. A shunt transmitter can use either a 5-volt or 12-volt supply and has low EMI but uses about twice the current of a nonshunt configuration that must use a 12-volt supply. Receive circuits can be analog or digital. The analog version has a higher bandwidth, less delay, and less pulse-width distortion, but it is more expensive. The digital receiver has a lower bandwidth and is less sensitive but is cheaper.

In-Car Wireless Communication?

Wireless (RF) communication could be used within a vehicle for serial data communications. Wireless technology is already used for remote keyless entry security systems and tire pressure warning circuits.

Obviously, the power would have to be kept low so that it does not interfere with the rest of the vehicle electronics or the vehicle in the next lane! Unfortunately, low power would make it susceptible to interference, especially from the vehicle's ignition system and from radio transmission towers. Many frequencies, such as AM, FM, cell phone, and so forth, would be off-limits for use.

Wireless could give the ultimate in wire count reduction and weight savings. Space would be conserved via smaller connectors at the nodes of the network. This topic is explored further in Chapter 16.

Summary of Transmission Media

Media in automotive use are not very fancy. But like microprocessor usage, the technology is often kept low so as to be robust and economical. Figure 7.5 is a chart of various media that shows distance versus speed.

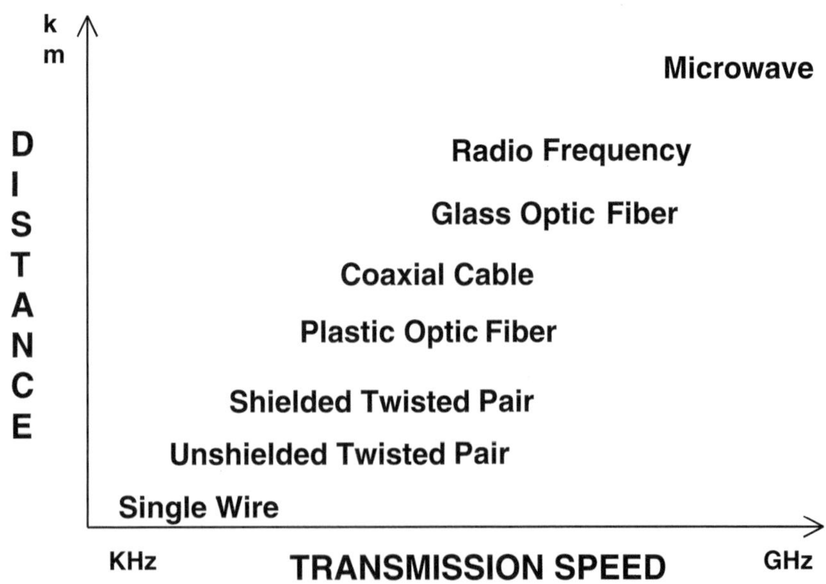

Fig. 7.5 Distance versus speed for various media.

References

[1] Cirillo, J., Jennings, K., Lynn, M., and Steele, R. "Physical Media Issues for High Speed Vehicle Networks." SAE Paper No. 941656, Society of Automotive Engineers, Warrendale, PA, 1994.

[2] Yumoto, N., et. al. "Optical Data Link for Multiplex Wiring." SAE Paper No. 830320, Society of Automotive Engineers, Warrendale, PA, 1983.

[3] Matsuzaki, Y., and Baba, K. "Development of Multiplex Wiring System with Optical Data Link for Automobiles." SAE Paper No. 840492, Society of Automotive Engineers, Warrendale, PA, 1984.

[4] Sawai, T., and Kawamura, S. "Receiver IC for Fiber-Optic Data Links." SAE Paper No. 870550, Society of Automotive Engineers, Warrendale, PA, 1987.

[5] Messuri, D., Miller, G., and Steele, R. "A Fiber Optic Connection System Designed for Automotive Applications." SAE Paper No. 890202, Society of Automotive Engineers, Warrendale, PA, 1989.

[6] Lefkowitz, R., and Coden, F. "Implementation of a Passive Star Based Fiber Optic Network for Full Vehicle Control." SAE Paper No. 890203, Society of Automotive Engineers, Warrendale, PA, 1989.

[7] Du Toit, D. "A Passively Tapped Linear Fiber Optic Data Bus." SAE Paper No. 890204, Society of Automotive Engineers, Warrendale, PA, 1989.

[8] Sawai, T., Sakamoto, F., and Hayashi, K. "Fiber-Optic Data Link for Vehicle Navigation." SAE Paper No. 900626, Society of Automotive Engineers, Warrendale, PA, 1990.

[9] Liello, L., Miller, G., and Steele, R. "A Passive Fiber Optic Network Designed for Automotive Applications." SAE Paper No. 9006271, Society of Automotive Engineers, Warrendale, PA, 1990.

[10] Yuuki, H., Ito, T., and Sugimoto, T. "Optical Star Coupler for Plastic Fiber." SAE Paper No. 910876, Society of Automotive Engineers, Warrendale, PA, 1991.

[11] Yonemura, R., Sawai, T., Kobayasi, N., and Kishimoto, K. " Passive Star Type Optical Communication Network for Vehicle." SAE Paper No. 930439, Society of Automotive Engineers, Warrendale, PA, 1993.

[12] Hetzel, H., and Schopp, P. "Advanced Entertainment and Communication Networking Based on Fiber Optics for Vehicle Applications." SAE Paper No. 940795, Society of Automotive Engineers, Warrendale, PA, 1994.

[13] Kawamura, S., Kishimoto, K., and Yuuki, H. "Development of an Optical Link for Automobiles." SAE Paper No. 940796, Society of Automotive Engineers, Warrendale, PA, 1994.

[14] Cirillo, J., Di Liello, L., Jennings, K., and Steele, R. "A Connectorized Passive Optical Star for Automotive Networking Applications." SAE Paper No. 940798, Society of Automotive Engineers, Warrendale, PA, 1994.

[15] Schreiter, G., and Engel, A. "New Types of Connectors for Plastic Optical Fibres for Vehicle Network Applications." SAE Paper No. 940799, Society of Automotive Engineers, Warrendale, PA, 1994.

[16] Daon, Q., Lynn, M., and Steele, R. "Diagnostic Procedures for Passive Optical Star Vehicle Networks." SAE Paper No. 940800, Society of Automotive Engineers, Warrendale, PA, 1994.

[17] Van Woesik, T., Post, J., and Kokken, J. "New Design of N*N Coupler and Connectors for Plastic Optical Fibres." SAE Paper No. 940801, Society of Automotive Engineers, Warrendale, PA, 1994.

Chapter 8

The Vehicle Level

Introduction

The vehicle level is where multiplexing realizes its potential. This is a high level where the individual nodes can be treated as "black boxes" and their contents ignored. The inner functionality of nodes is discussed in succeeding chapters at the ECU level and the IC level.

Figure 8.1 shows the level of implementation that this chapter explores. The vehicle is shaded to indicate that we are looking at the "big picture."

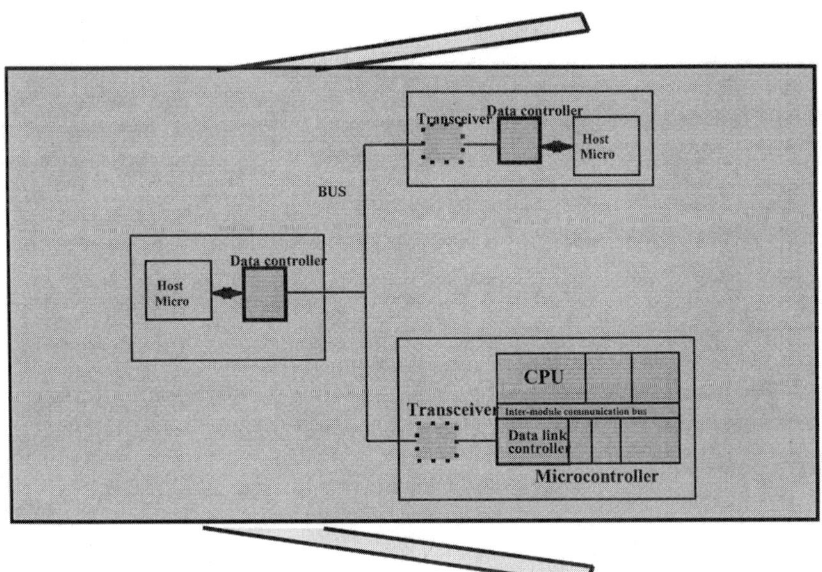

Fig. 8.1 Vehicle level.

Topologies

First, a brief review of possible vehicle network topologies is in order. These are "logical" configurations without much regard to the actual physical media used. Most topologies may be implemented with one wire, two wires, or fiber optics.

The star (seen in Fig. 8.2) is most often used for master–slave types of protocols where a centralized node coordinates traffic among the other nodes. If there is a center node, all messages have to go through this node and may be repeated by the master node.

No center or master node is present in other implementations of the star. Instead, the legs of the network go to a bus bar or some other type of connector. This is supposed to facilitate diagnosis but really negates the benefit of a serial data link. The arrangement, however, has been used on several automobiles in production, such as the 1998 Corvette (see Fig. 8.3).

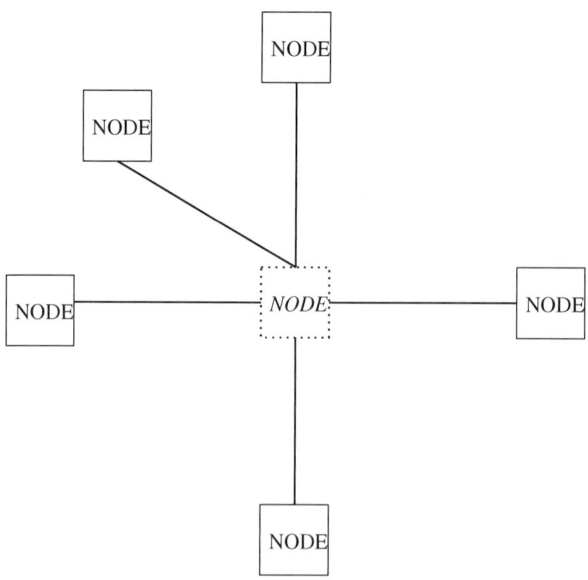

Fig. 8.2 Star topology.

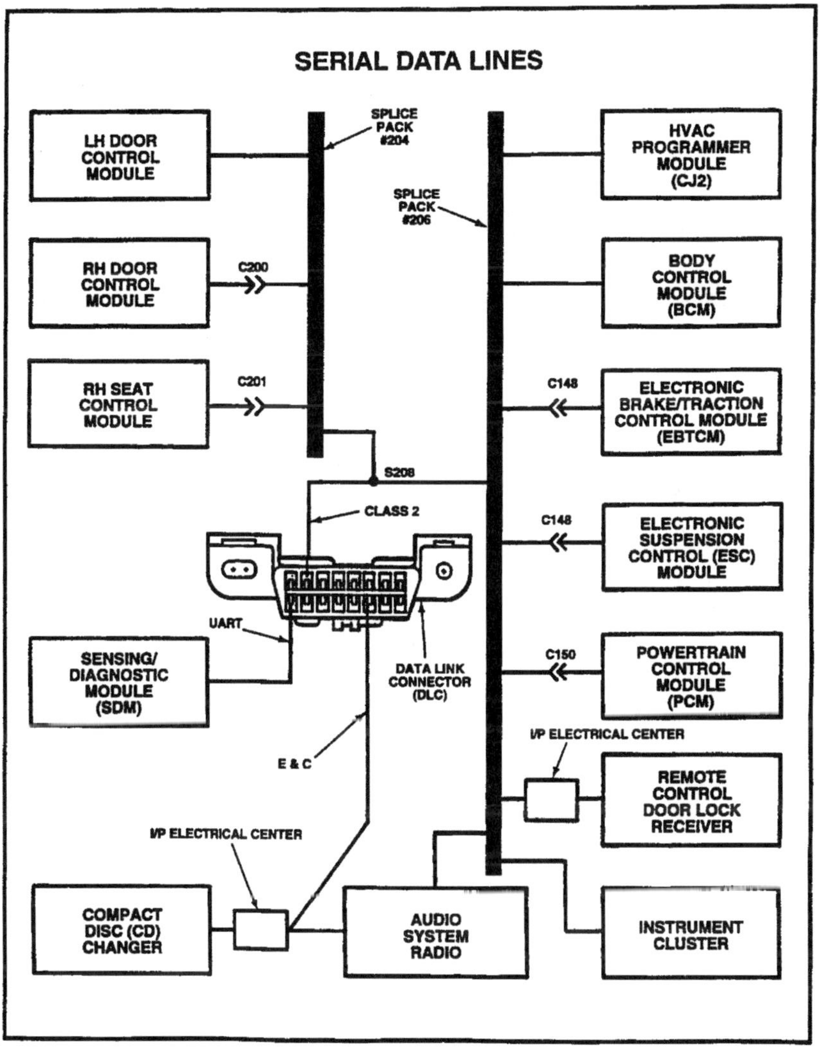

Fig. 8.3 1998 Corvette basic network topology.

The media that can be used for a star are single wire or fiber optic. Differential twisted pair would not generally lend itself to this type of topology because of difficulty in terminating the "ends" of the network.

An attractive feature of the star structure is its ability to tolerate an open or short in one of its legs without losing communication among the other nodes (if the center node is present). A major disadvantage is that all messages must pass through the center node (if present).

The ring (Fig. 8.4) is a common structure and is actually a bus (Fig. 8.9) whose ends are tied together.

The ring structure provides a redundant electrical path for messages, if there is an open somewhere in the network. A shorted wire may also be tolerated, if each node has two transceivers (Fig. 8.5).

The only copper medium suitable for rings is single wire. Fiber optics is another possibility and is used with protocols such as MOST.

In practice, the dual transceiver approach is never used because of the extra hardware and complexity. The trade-off between this and extra reliability does not warrant the two transmitters. It is worth noting that many people refer to this as true "daisy-chaining," meaning that a message must be retransmitted from node to node.

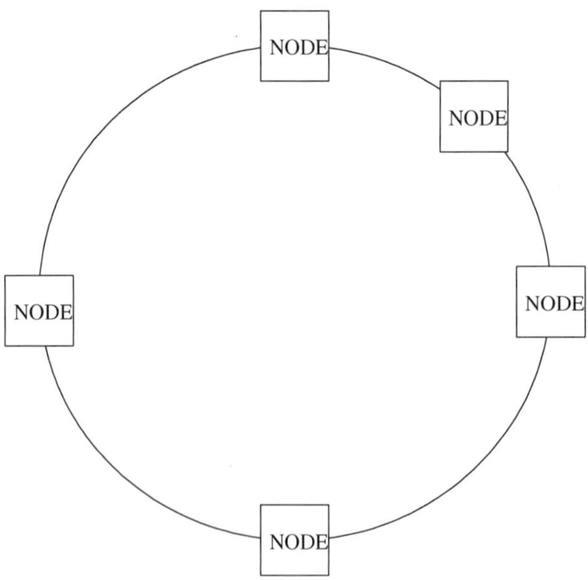

Fig. 8.4 Ring topology.

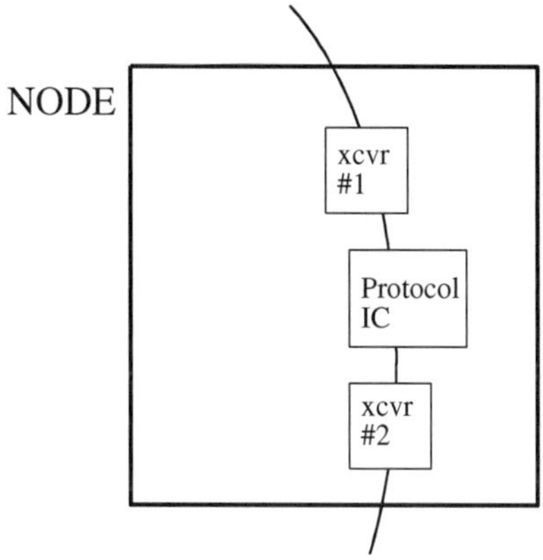

Fig. 8.5 Node detail for dual transceiver configuration.

Without the dual transceivers, a short will bring down the entire link because there is only a single path from the transmitter/receiver to the channel. The usual way of attaching a node to the network is by daisy-chaining the medium from one node to the next. In most cases, the splice is within the node—on the circuit board (Fig. 8.6). However, if such a node is unplugged from the network, it leaves an "open" which could affect operation of the other nodes.

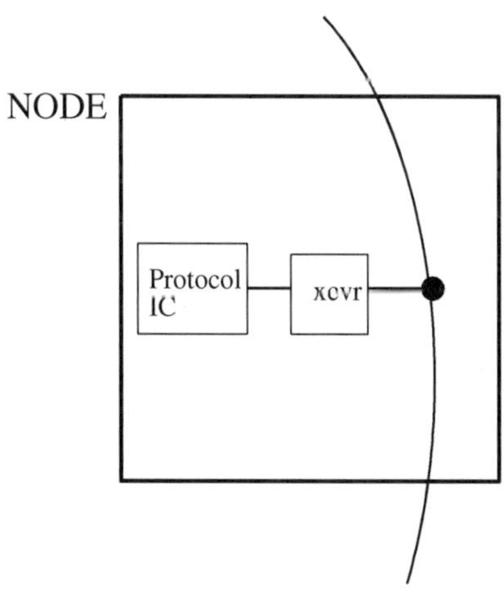

Fig. 8.6 Node detail for single transceiver configuration.

It is becoming more common for vehicle manufacturers to allow splices in wiring harnesses. This is cheaper and makes it easier to assemble the vehicle (Fig. 8.7). This approach also needs only a single set of connector pins, whereas the previous configurations required twice the number of serial data connector pins per box to connect the nodes.

A ring is somewhat more difficult to route because the ends of the bus must be brought together. This may necessitate long wire(s).

The tree is a broad category of network structures that usually implies multiple levels. Figure 8.8 illustrates what could be called a three-level bus structure. The three nodes at the left may be closely related nodes in proximity or functionality. If a node is present at the dashed box, the three-node "subbus" may even use a protocol different from that used by the other nodes. Similarly, other groups of nodes may be clustered because of geographic location or function and may possibly be separated by a gateway to other nodes.

A tree is a common vehicle-level network structure. Usually, routing and connector constraints make it necessary to have at least the appearance of multiple bus structures.

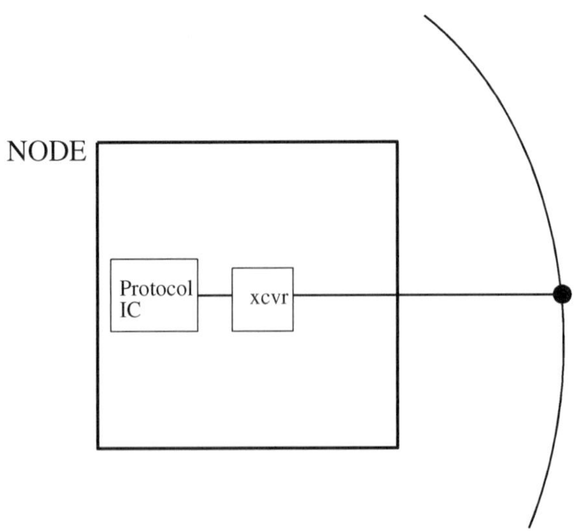

Fig. 8.7 Node detail for single transceiver with harness splice.

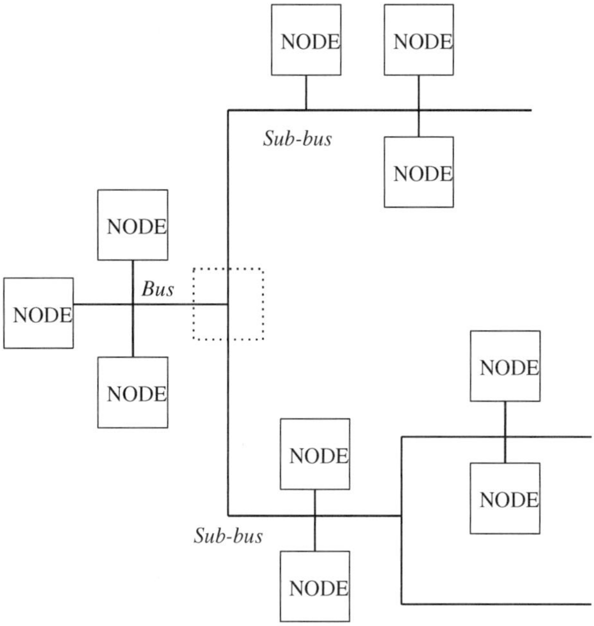

Fig. 8.8 Tree topology.

Single wire, twisted pair, and fiber optics may be used to implement tree structures because the definition is so broad.

A tree gives great flexibility in routing, but, as in a bus structure, a short or open may take much, or all, of the network off-line. It may also be difficult to diagnose if intervening gateways are present.

The point-to-point topology, shown in Fig. 8.9, is rather trivial but deserves mention because it is often used. It is a simple configuration that allows using a noncomplicated protocol (such as a UART). One of the nodes is typically a sort of master node such as a tester or programmer. The other node is commonly the engine controller such as in ISO-9141 protocol applications.

A benefit of point-to-point is that there is a minimum of connections and no splices. Disadvantages are that an open bus wire takes down the entire network, and the topology can obviously be used for only two-node applications.

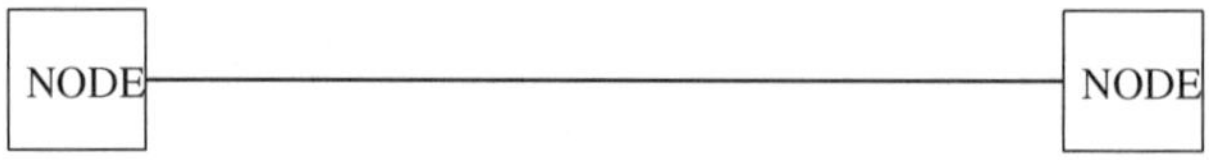

Fig. 8.9 Point-to-point topology.

The bus (Fig. 8.10) is the primary automotive multiplex topology. Typically, though, it is more of an ideal than a realizable structure.

A bus is simple, has a low number of tie points and connections, is straightforward to route, and has no implied order of nodes. It does not, however, have the benefit of an alternative current path like a ring. A broken or shorted wire may well take down the entire network.

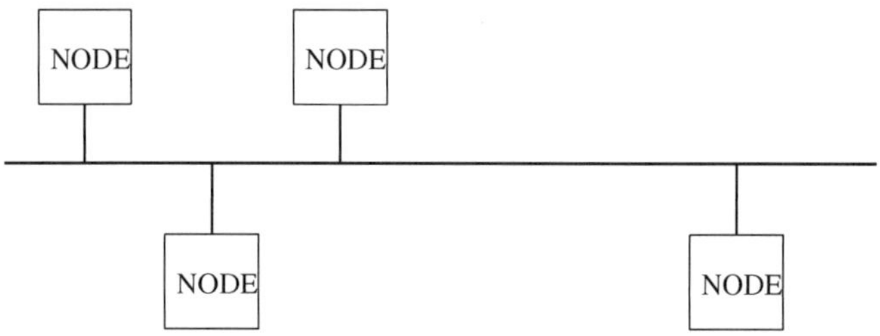

Fig. 8.10 Bus topology.

Figure 8.11 illustrates a key point of vehicle-level multiplexing in that more than one network is usually necessary. More than one protocol is usually required—sometimes as many as four or more [1]. "One size fits all" does not work. There may be multiple networks with the same protocol, such as CAN, because it is somewhat easier and cheaper to support a lower speed bus than a high-speed one. Lower-cost CPUs, less software, and so forth are necessary to implement node interfaces to a 100 Kb/s CAN network than at 500 Kb/s.

There are good reasons for certain applications such as airbags and entertainment (mobile media) to have their own communication bus. An airbag network may have as many as 64 nodes when you count all of the sensors, belt

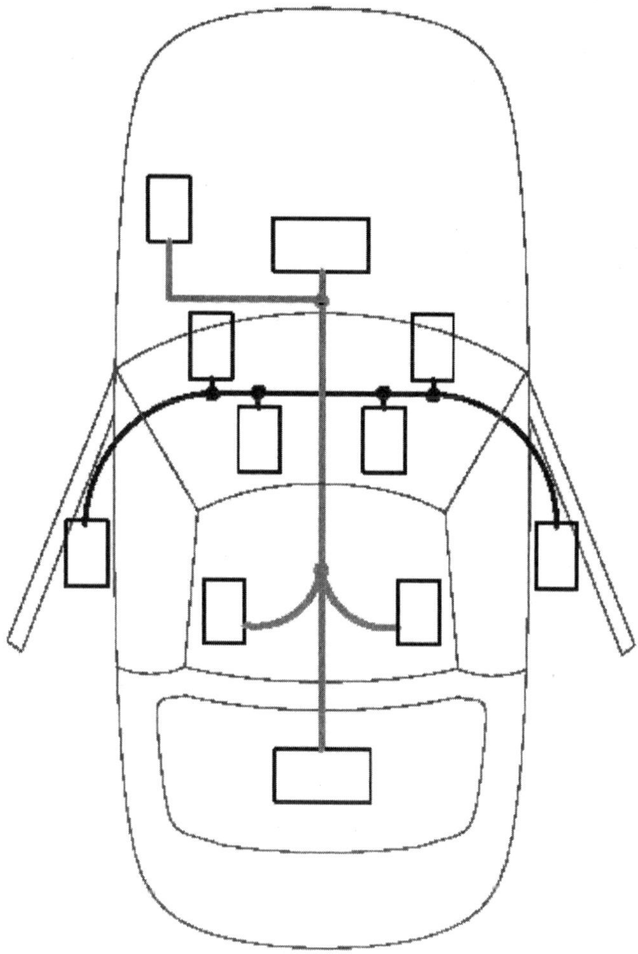

Fig. 8.11 Example of multiple-networked vehicle.

pretensioners, squibs, and so forth. This network needs to be available literally on a moment's notice. It cannot afford to support many nonairbag messages. Likewise, a modern entertainment network supports real-time, high-speed speaker data and video streams. It is probably a 100 Mb/s (or faster) isochronous link with no room to carry powertrain control or blower fan speed messages.

One of the latest additions to the vehicle network family is real-time critical control functions for "X-by-wire" applications. This encompasses brake-by-wire, throttle-by-wire, and steer-by-wire. The time-triggered protocol (TTP) is an example of a protocol that will be used for this type of bus. Such protocols are discussed further in Chapter 16.

Network Design Issues

It is a good idea to use the same physical layer circuit at each and every node on the network. For example, if one is implementing a two-wire CAN bus, then use the same transceiver IC and a common "building block" for the physical layer interface of each node.

Requiring a common physical layer interface avoids many problems during systems integration. Some of these problems would be from using parts that had different V_{oh} levels, different comparator input switching thresholds, and so forth. The differences may be subtle and still within the applicable physical layer specification (i.e., ISO 11898) but are still important. These differences may show up, especially during bus fault and/or transient conditions. The reaction of each node electrically to bus events should be the same.

DaimlerChrysler mandates a common physical layer for its J1850 applications. For several model years, it was the Harris HIP7020. Now, it is usually the Philips AU5780. GM chose not to require a common J1850 physical layer interface, and tough systems integration challenges resulted for some vehicle models.

It must be remembered that multiplexing is a tool for use in vehicle system design. Multiplexing can be used properly or improperly. It does no good if the right partitioning of functions is not done correctly. The consequences of bad vehicle electronic architecture can be more severe when using multiplexing [2].

Development Tools

Development tools are important in the entire realm of automotive multiplexing—from the vehicle level down to the IC level.

These tools are typically used at a bench, prevehicle, environment to debug one or more nodes. Development tools are indispensable for systems integration, either on the bench or in a prototype vehicle.

There are many sources for these tools, including Dearborn Group Technology, Inc. in Farmington Hills, Michigan (248-488-2080), and Vector CANTech in Novi, Michigan (248-449-9290).

Service Tools

Service tools, as opposed to development tools, are those typically used at car dealerships. They are available only after the vehicle is in production. These tools usually must encompass a wide range of serial data link protocols. For example, the U.S. "Big Three" service tools have, or soon will have, capability for UART, J1850, CAN (one-wire and two-wire), and other protocols—all in one tool.

Major manufacturers of service tools include OTC, Balco, MPSI, and Hewlett Packard. Knowledge-based tools may be coming in the future [3].

EMC Issues

Eventually, serial data networks will encompass electric vehicles as well as internal combustion applications. Studies have shown that protocols such as CAN and J1850 will be well suited to these harsh environments [4].

Vehicle Level Testing

Carmakers perform vehicle tests at their facilities. These tests are usually based on SAE or other world standards, but not exactly. Examples of tests would be the GM GMW-31xx series of evaluation procedures that are not public documents. Some of them use a large "screen room" and/or anechoic chamber into which the vehicle is driven. Instead of GM GMW specifications, the older GM-91xx tests will be referenced in the text later in this book.

Potential Results

Implementers of multiplexing have reported savings of 20% in wiring harness complexity and 10% cost reduction. Reliability and flexibility are also greatly improved [5].

References

[1] Emaus, B. "Aspects and Issues of Multiple Vehicle Networks." SAE Paper No. 950293, Society of Automotive Engineers, Warrendale, PA, 1995.

[2] Lindgren, M., and Persson, P. "The Multiplexed Vehicle—A Maintenance Time Bomb?" SAE Paper No. 960391, Society of Automotive Engineers, Warrendale, PA, 1996.

[3] Rizzoni, G., and Soliman, A. "A Survey of Automotive Diagnostic Equipment and Procedures." SAE Paper No. 930769, Society of Automotive Engineers, Warrendale, PA, 1993.

[4] Powers, C., and Huettl, T. "Using Existing Multiplex Communication Technology to Implement an Electric Vehicle Communication Network." SAE Paper No. 951887, Society of Automotive Engineers, Warrendale, PA, 1995.

[5] Voss, W., and Butscher, K. "In-Vehicle Data Bus Systems—The Key for New Concepts in Comfort and Convenience Electronics." SAE Paper No. 960121, Society of Automotive Engineers, Warrendale, PA, 1996.

Chapter 9

The Electronic Control Unit (ECU) Level

Introduction

Several electronic control unit (ECU), or "box" level, issues are discussed in this chapter. The first is the advantages of integrating functions. Another is fault tolerance. Some network design subjects are presented that take place at the ECU level. The final topic is design metrics.

Figure 9.1 shows the level of implementation this chapter explores. At this level, we look at each node as a "black box" that must correctly communicate with all of the other nodes.

Integrated Controllers

As discussed in Chapter 3, electronic functionality must be partitioned within the vehicle. Usually, this has followed natural boundaries such as ABS having its own separate box. This, however, does not have to be the case. The ABS control could be integrated within the engine control box. GM has already done this with its line of vehicle control module (VCM) ECUs for light-duty trucks.

A major advantage of integrating functions in a lesser number of controllers is lower cost by reducing the duplication of functions. One CPU, one voltage regulator, and so forth can conceivably do the work of two or more functions.

Integration can be taken to the extreme where there is only one ECU per vehicle, a " PC in the trunk," if you will. Several consumer computer-oriented companies such as IBM and Microsoft would like nothing better than becoming suppliers of electronics to the "office on wheels."

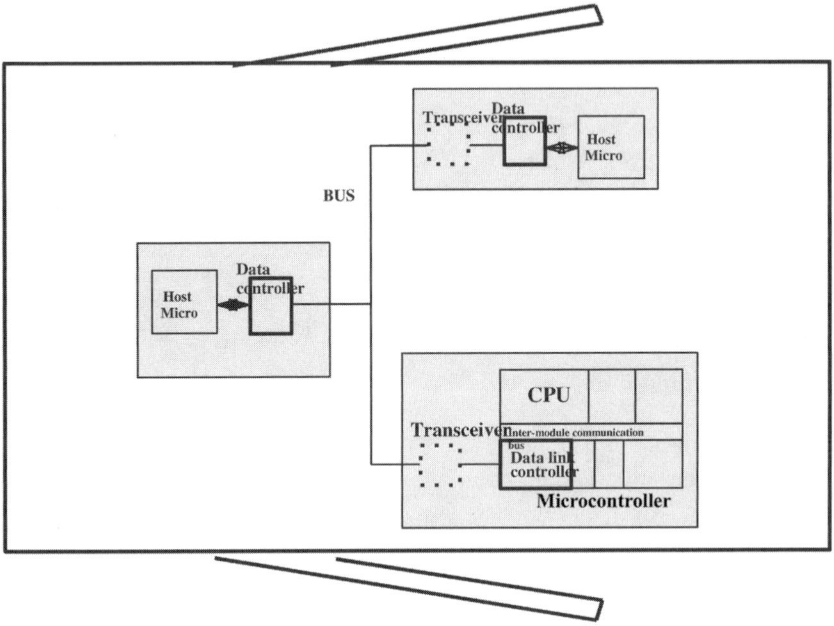

Fig. 9.1 ECU (node) level.

The single controller scheme could be listed as a competitor to multiplexing because there is no network but only one node. The major disadvantages of a single, unified, controller are large size, large number of wires connecting to it, complexity, low fault tolerance, and difficulty in standardizing its functions. The controller would be very inflexible to changes in vehicle content.

Fault tolerance is actually a big area of discussion among automotive electronic engineers. It is hard to quantify because it is affected by the assumptions about noise, shorts, and other environmental factors to which you assume the system is exposed. For example, an argument can be made that one controller doing two separate functions is more reliable than two controllers each doing one of the functions. The single controller is a smaller target for noise, and there are fewer components to go bad. Even if redundant controllers were used (two controllers doing the same function), both are likely to be exposed to the same noise, transients, and faults.

The preceding discussion is also applicable to any debate over one-wire versus two-wire serial data links. A one-wire bus essentially integrates a two-wire bus, although it does open new problems regarding EMC (Chapter 11).

Several network design issues need to be raised. The first is scheduled versus on-demand messaging. Scheduled, or periodic, messaging is network messaging sent at a fixed interval. This interval may be 1 μs (or less), 1 ms, 10 ms, 2 s, and so forth. Usually, these messages are the result of information, such as engine speed, vehicle speed, and so forth, that changes at some constant rate. These kinds of data are typically found in Class B and Class C systems.

On-demand, or event-driven, messaging consists of messages sent when the data change or are requested by another node. An example is a button pushed to engage cruise control, which would cause a message to be sent. There are usually minimum and maximum update rates for these messages. Event-driven messaging is most common in Class A and Class B systems.

Systems integration is very important in proper ECU design and implementation. Because each node individually performs its serial data functions correctly does not mean it will perform properly when connected to the network. The many reasons for this involve both hardware and software. A typical hardware problem is having slightly different physical layer (analog) transceiver interfaces on the network. Although their specifications may seem compatible, there will be differences in the way they react to transients, faults, noise, and so forth. This may cause varying reaction at the node's outputs and possibly have an effect noticeable by the vehicle operator. Within most of DaimlerChrysler's J1850 usage, one and only one IC is allowable to perform the transceiver function within ECUs. This is a good way to force compatibility among ECU suppliers but offers limited flexibility in the designs.

Typical software issues that affect systems integration involve handling certain message events. Some protocol handling software designs treat normal and error messages differently among the nodes. This, as with the hardware issues, usually causes some network behavior that the operator may notice. This behavior may not be directly involved with the node that exhibits the problem. For example, if a "Break" symbol occurs on a J1850 bus, it will clear out the transmit buffers of all nodes. Most nodes will handle this gracefully and reload any message that was in progress during the Break. Some node, however, may choose to ignore the Break and may never even know it happened. The problem is that if it was transmitting a message at the time of the Break, some other node may have been awaiting the message. Because it never comes, the other node does some error processing and may even store a trouble code or exercise a system reset. Therefore, a difference in handler software at a transmission controller could cause anomalous behavior at the instrument cluster node.

A common hardware and software handling strategy for the network can prevent these problems. It is very likely that nodes will come from several different suppliers, but they all have to be designed to the same specification. This is especially true for fault processing and handling marginal network behavior such as a resistive load that is right at the absolute minimum level. There must be a generic software handling specification to handle various message strategy options, IC options, and fault conditions but yet not force undue burden in ROM allocation.

When ECUs are designed, it should be to well-thought-out specifications that are the result of functional partitioning exercises. The questions of which protocols to use, how many functions will be handled, and which nodes will get which functions must have been answered by the vehicle electronic architect.

Unexpected Message Delays

Another item that can wreak havoc with network behavior is misallocation of message priority on a contention (arbitrated) link. Consider the following example of a set of messages that needs to be transmitted onto a network.

Message	Prority:
Vehicle_Speed	5
Knock	6
rpm	5
Oil_Temp	4

These messages have various priority levels from four to six; four is the highest priority (assume a zero-dominant system). Each message comes from a

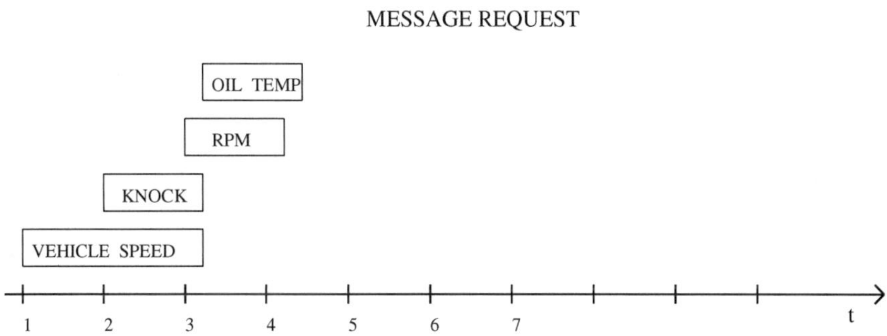

Fig. 9.2 Example of priority message request order.

different node and has an arbitrary length. Figure 9.2 shows the order for queuing these messages to transmit them onto the network. Vehicle Speed is first and actually takes possession of the channel. During this time, the Knock message is ready to transmit at some other node but cannot because the bus is busy. Before Vehicle Speed finishes transmission, two other messages (rpm and Oil Temp) are queued for transmission at two other nodes.

When Vehicle Speed finally finishes transmission and the bus is idle, there are now three messages that will arbitrate for control of the bus. The one with the highest priority, Oil Temp, will be the next message to transmit completely. Note that Oil Temp was the last to queue up among the three but is the first to transmit. When Oil Temp completes, rpm and Knock will try again to gain the bus and will contend. rpm wins out because it has a higher priority than Knock. When Knock tries again after rpm finishes, it will finally get through. This scenario is illustrated in Fig. 9.3.

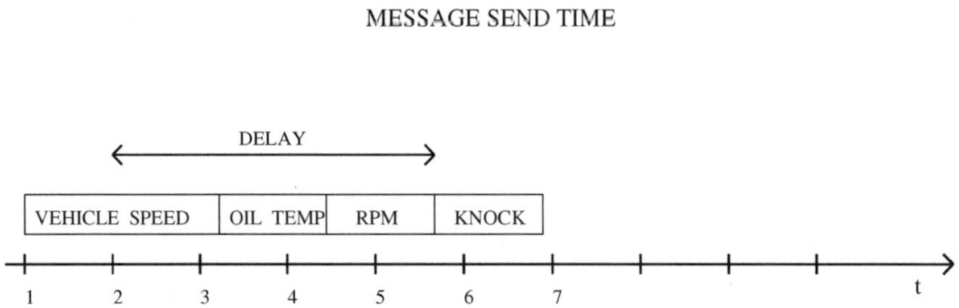

Fig. 9.3 Example of actual priority message transmit order.

Notice in this example that even though Knock was ready to go before Oil Temp and rpm, it ended up transmitting last because of its priority level. Maybe this is OK and will have no ill effects. Maybe the network designer will need to run another round of simulations with a higher priority level for the Knock message.

Message Synchronization

In general, contention is not a good thing. Contention, or arbitration, results in three forms of source synchronization—clustered, random, and balanced [1]. Many network designers deliberately set the maximum allowed bus throughput very low, so that there are few arbitrations. Sometimes this is hard to do, for example, at "key-on." Especially during an ignition cycle, or key-on, a situation called "clustered synchronization" may occur. Clustered synchronization occurs when many messages from many different nodes are ready to be transmitted at about the same time. Some messages will spend a lot of time waiting for retries because of the high network use.

Although mostly unavoidable, clustered synchronization can be eased by having a node wait as long as possible before transmitting. The ECU should listen to the bus for a second or so to try and fulfill its initialization list of parameters before requesting these items.

The normal condition for a contention link is "random synchronization." This is usual when messages transmit at random intervals and probably rarely collide. Sometimes, though, messages from different nodes may be on approximately the same time-out period and almost always collide. For example, the powertrain control module (PCM) may send vehicle speed on a 5-ms loop period, whereas the transmission controller sends torque converter lockup status every 10 ms. If the timers at these nodes somehow synchronize, then the PCM vehicle speed message will collide with the transmission message every other time. The way to avoid this is to use a variety of loop times at each of the nodes.

"Balanced synchronization" is a forced condition where collisions almost never happen. It is an ideal state when bus traffic is relatively low and nodes take turns transmitting. The transmission points for each node are interleaved, so that messages do not line up. This is very difficult to achieve. A possible solution is to imitate the Ethernet protocol. When a collision occurs, each of the colliding nodes can vary its loop time a little bit or wait slightly next time if the data were event driven. Ideally, this will decrease the number of collisions.

Importance of Delay

When analyzing the delay properties of networks, the worst case delay is more important than the average delay for a message, especially for Class C, real-time systems. This gets us back to one of the primary facts of serial data links. Master–slave and token–slot protocols have fair average and worst case delays, and they are easily predictable. Contention-type protocols can have terrible average and worst delays, unless message priorities are assigned correctly. In this case, the arbitrated network can have excellent average and worst case delays—better than master–slave and token–slot systems.

Performance Metrics

There are several metrics that network designers use to benchmark an ECU on a network. The first of these is system or ECU cost.

System Cost

System cost refers to what the entire node costs—hardware, software, and all. This is generally fixed and easy to quantify. It is useful in comparing the cost of one node to another, but obviously there are other factors (e.g., the price of a PCM is going to be more than a mirror-control module). Another metric is functional latency.

Functional Latency

Functional latency is a measure of how long it takes to cause a specific function or action. Not all functions have a latency requirement, but it is very important for those that do. Latency can be measured many different ways. One of the most insightful is indicated in the formula below for message delay, t_d, from message request for transmission to action:

$$T_D = L_I/R_T + D/C + T_{LI} + T_{LO}$$

where

T_D = total message delay from request to send until completion, in seconds

L_I = length of message I, in bits

R_T = message transfer rate, in bits per second

D = length of the network, in meters

C = message propagation speed = approximately 2×10^8 m/s (2/3 the speed of light)

T_{LI} = message latency from request to beginning of transmission, in seconds

T_{LO} = message latency from receipt to completion of action, in seconds

Figure 9.4 shows these components graphically. This depiction of latency is good because it encompasses all portions of the delay from the time an action is requested to when it actually gets done. It is not popular because it is hard to measure a couple of the components, namely, T_{LI} and T_{LO}. So often, they are ignored.

Cost per Function

Cost per function assigns a price to each function, based primarily on importance.

Communication Overhead

Communication overhead encompasses the worst case message latency and thus the functional latency. This involves how much time it takes to transfer a piece of data.

Communication Link Throughput Efficiency

This metric predicts the average data throughput and overhead from nondata bits. Code density is a metric relating to software load to handle the network functions.

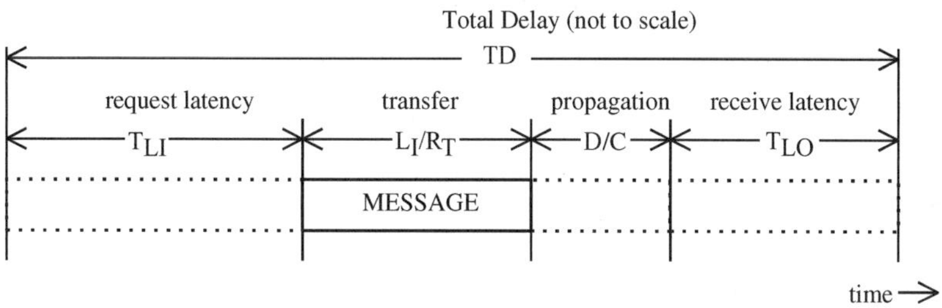

Fig. 9.4 Functional latency time line.

Code Density

Code density is a measure of stored instruction bytes per function.

System Overhead

System overhead is the number of CPU cycles needed to support the network.

Message Data Ratio

Message data ratio is a software metric. This is the ratio of data link data to all the other data handled by the ECU.

Fragmentation

Fragmentation is the cost of unimplemented node resources. It results from inefficient system partitioning.

Mechanizations

As an example of mechanizations and hardware partitioning, the Class 2 protocol is implemented in several ways at the ECU level. One example of using a peripheral chip, such as the data link controller (DLC) (Motorola MC68HC58) IC, in a microcomputer system is shown in Fig. 9.5. The DLC IC (Motorola

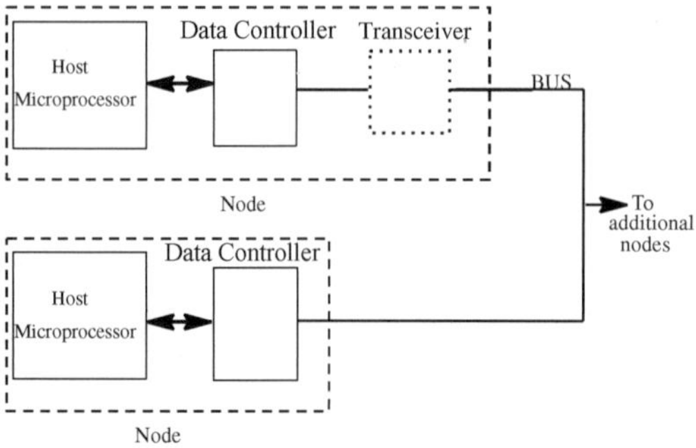

Fig. 9.5 Block diagram of a Class 2 node using all peripheral circuitry.

MC68HC58) contains the transceiver, but some other implementations do not and will thus require a separate analog driver IC or discrete circuit.

Another approach is to up-integrate the logic circuitry (and possibly the transceiver) onboard the microprocessor, as shown in Fig. 9.6.

The Class 2 protocol allows for as many as 32 nodes, including a diagnostic or scan tool plugged into the assembly line diagnostic link (ALDL) connector.

Built-In or Offboard Transceiver

The analog (driver/receiver) portion of Class 2 is integrated in some solutions (DLC, message data link controller [MDLC], PRISM, for example) and a

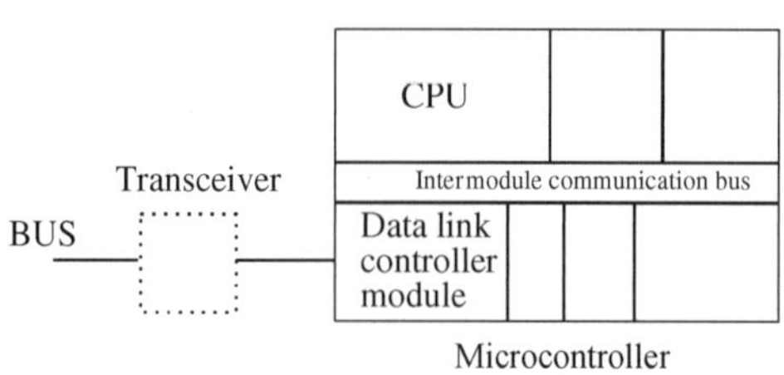

Fig. 9.6 Block diagram of a Class 2 node using up-integrated logic.

separate IC on others (byte data link controller [BDLC], data link controller module [DLCM], for example). The industry position used to be that the transceiver should be packaged with the digital circuitry, if at all possible. Today, this has changed. Now, because of very small digital process techniques, it usually does not make sense to bundle the analog circuitry for the transceiver with the digital portion. What is happening is that IC manufacturers are bundling various analog functions together (e.g., transceiver(s), power supply(ies), motor drivers) onto a separate "power" chip.

Bus Line Routing

The bus pin or pins should be as close as possible to the external box connector to prevent contamination by other noise sources within the system. Also, route the bus traces (inside and outside of the box) as far as possible from other traces containing sources of noise such as solenoid drivers, and so forth.

If multiple bus lines are used in the product via daisy-chaining, they should be adjacent within the product connector. This eliminates one fault problem, that of adjacent pins shorting together. In this case, it would have no effect.

Bus Loading

GM's Class 2 bus is a good illustration of the typical "loading" that must be distributed over the network. The summary on bus loading for a Class 2 node is that one node will have a total of 3200–3300 pF and 1.5 kΩ on the bus line to the load line. All other nodes must have 470 pF to ground and 10.7 kΩ between bus and load.

Note that one Class 2 node in the vehicle (the ECM or PCM is recommended) must have a heavier bus pin loading, so that there is a smaller differential between the total network loads of 2 nodes versus 32 nodes. Also note that if 26 nodes or more are to be used, the heavier loading must not be present because this would violate the maximum capacitance and minimum resistance allowed by J1850. Thus, all 26 or more nodes in a network should have "non-PCM node" load values. The battery voltage pin of the transceiver is usually attached to the permanent "unswitched" battery supply to take advantage of the better transient protection generally found on this circuit rather than "switched" battery, or ignition.

If other components, such as a transient suppression Zener diode(s) or a varistor, must be added externally, the loading should be compensated for accordingly because these components may have substantial capacitances of their own.

Local Loss of Power and Ground

It is imperative that a local (at the box or ECU level) loss of +5 V or battery voltage, or both, not disrupt the rest of the bus. Again, the Class 2 bus is used as an example. The requirement is usually guaranteed by placing a specification of ±1 mA maximum leakage current on the bus pin to ensure that V_{il} (1.5 V) of the receivers cannot be met by one node that is leaking current onto a minimally or maximally loaded bus. For an illustration of the leakage path from a primary node losing ground, see Fig. 9.7; for a secondary node losing ground, see Fig. 9.8.

Any loss of ground for a product, system, or box must be tested by pulling *all* ground connections in the circuit containing the Class 2 hardware, especially the ground(s) for the power supply that feeds the Class 2 IC/module.

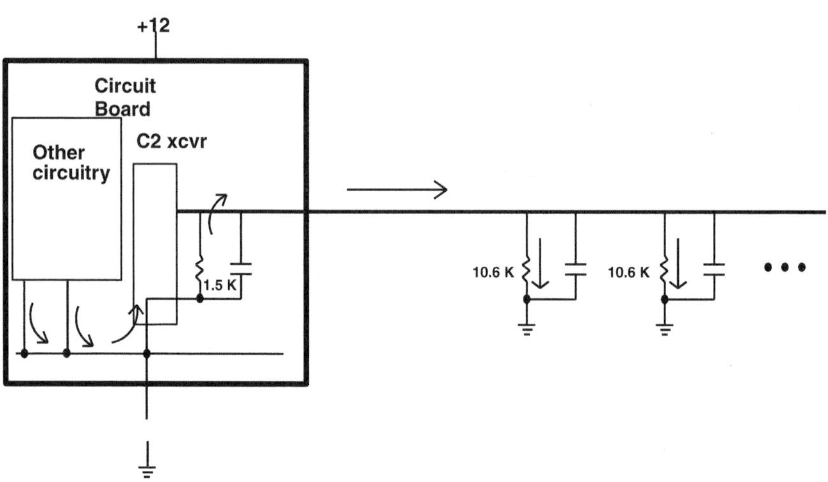

Fig. 9.7 Loss of ground (primary load).

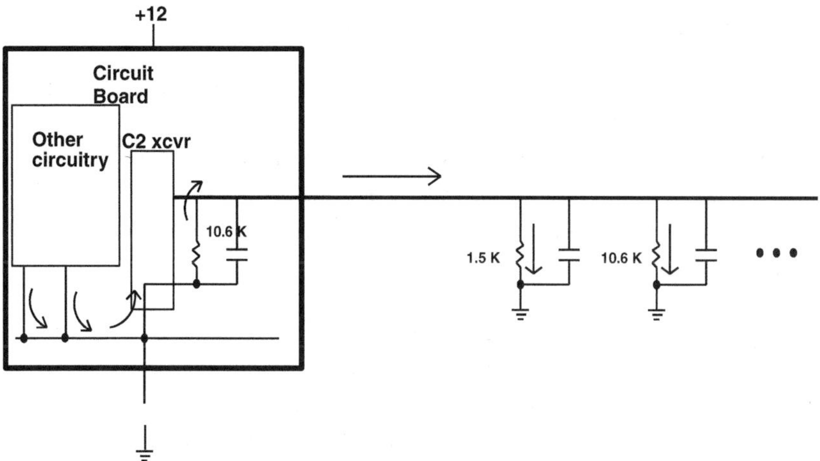

Fig. 9.8 Loss of ground (secondary load).

The method used to test for loss-of-ground performance at the building block level is to start with the Class 2 IC/module with only its required peripheral components. The lines Vbatt, Vcc, Vdd, Vssa, and Vssd are tied together and grounded. Then, the bus pin may be tied to –16 V, and the leakage out of the bus and load pins are added together and must be less than 1 mA out of the part.

If the loss of ground is tested on a network such as on a development vehicle, then it is important to note which node(s) have ground removed and the resulting load impedance for the nodes that retain ground. When ground is removed from a Class 2 node, you have then removed some of the resistive pulldown from the network and coincidentally changed the bus time constant. If ground is removed from the primary load (i.e., the PCM), it is possible that the remaining load violates the J1850 and Class 2 minimum load requirement for at least eight unit, or secondary, loads. This is important because most Class 2 devices are allowed to leak up to 1 mA from their bus pins during a loss-of-ground test. The following equation needs to be understood:

$$V_{off} = (0.001\ n)\ (R_{bus}) \tag{9.1}$$

where

V_{off}	=	bus bias voltage as a result of groundless nodes leaking current onto bus (V)
0.001	=	maximum current leakage from each node losing ground (A)
n	=	number of nodes losing ground
R_{bus}	=	parallel pull-down resistance of remaining nodes that still have ground (Ω)

If V_{off} exceeds approximately 3.8 V, then communication will be disrupted among the nodes that retain ground. The original intent of loss-of-ground testing of a Class 2 system is that communication would not be disrupted as long as ground was being pulled to one node at a time, and the remaining nodes still see a bus load between 300 and 1500 Ω. Any other situation must be analyzed case-by-case using Eq. (9.1).

After examining Figs. 9.7 and 9.8, you can see that the voltage bias produced on the bus by a node losing ground differs when the node is either a primary or secondary load. The voltage is greatest with the secondary load because the leakage will be through a 10.6 kΩ resistor, instead of the 1.5 kΩ that is at the primary load.

The method used to test for loss-of-power performance at the building block level is the same as that described in the previous section, except that the bus pin is tied to +16V, and the "leakage" current is measured going into the part.

EMC

When testing a product, the product should be tested in two configurations, minimum bus load and maximum bus load.

For example, in Class 2, a minimum of eight unit loads should be present on the bus. For a PCM test, this means having one additional unit load connected to the bus pin of the product. When testing a secondary loaded product (non-PCM), a lumped load equivalent to seven unit loads must be added.

For maximum load, 32 unit loads should be present on the bus. For a PCM test, this means adding a lumped impedance equal to 25 unit loads to the bus pin of the product. When testing a secondary loaded product (non-PCM), a lumped load equivalent to 31 unit loads must be added.

Bulk Current Injection (BCI)

Some Class 2 components require series impedance to enhance radiated transient immunity. The corresponding GM test is GM9112P.

Electrostatic Discharge (ESD)

Some customers of products using Class 2 have begun requiring the bus line to pass GM9109P/19P to a Class C level and ±15 kV. Not all ICs/module need external protection. *Note that the protection schemes described in the subsections below are not required in all cases and are used only as examples.* A modification of the usual test procedure is necessary to test for the damage that would occur (latent damage).

- Unidirectional Zener Diodes (Fig. 9.9)

  Two diodes are necessary to protect from positive and negative spikes, as well as prevent the Class 2 bus from being clamped to ground during a local loss-of-ground condition. If Zener diodes are added to a circuit, the load capacitance on the bus pin must be adjusted to compensate for the capacitance of the diode stack, which is approximately 235 pF. For a primary node, the capacitor should be changed to 2700 pF (or as close to

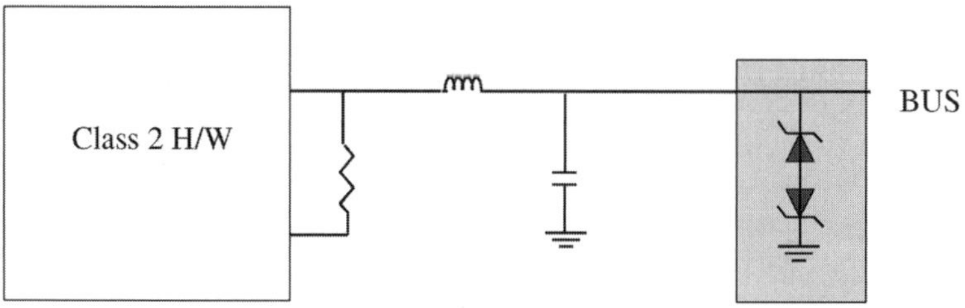

Fig. 9.9 Optional ESD protection on bus pin—two diodes.

3065 pF, if using multiple capacitors because of connector filtering, etc.). Secondary nodes must be changed to a 220-pF cap. Examples of diodes are General Instrument TPSMA16A (surface mount), General Instrument P4K16A (leaded), and Motorola P4SMA16AT3.

- Bidirectional Zener Diodes (Fig. 9.10)

  A bidirectional surge diode is available now to take the place of the two single parts. The load capacitance must be adjusted to compensate for the capacitance of the diode stack.

  General Instrument SMAJ15C (surface mount) and Motorola 1SMA15CAT3 are examples of bidirectional diodes.

  These diodes are functionally equivalent to a pair of the unidirectional diodes discussed previously. The unidirectional diode may still be useful for certain applications that "share" one of three diodes in protecting two different, one-wire data buses.

- Metal Oxide Varistor (MOV) (Fig. 9.11)

  Metal oxide varistors (MOVs) are the latest and best technology for ESD protection of serial data buses [2]. MOVs are generally smaller and cheaper than transient suppression diodes and do not suffer from their "wear-out" phenomena.

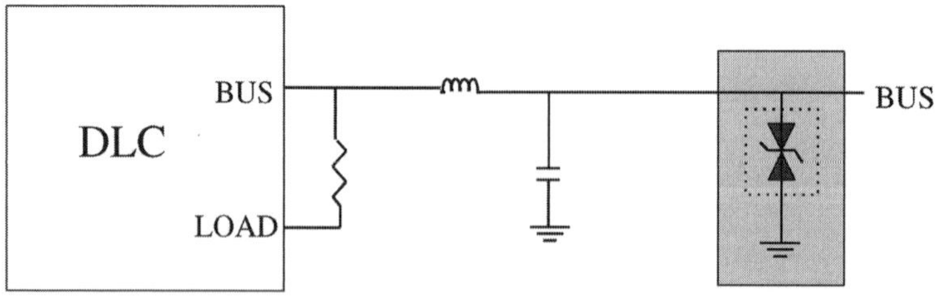

Fig. 9.10 Optional ESD protection on bus pin—bidirectional diode.

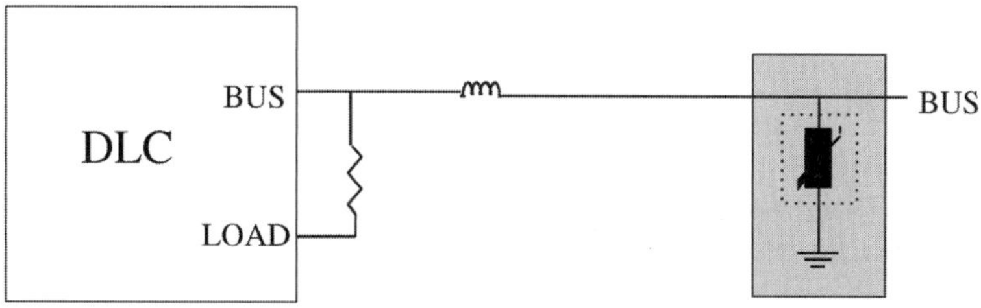

Fig. 9.11 Optional ESD protection on bus pin—metal oxide varistor (MOV).

A metal oxide varistor in an 0805 package is available now from AVX (an example is AVX VC080510A230). An added bonus of this method of protection is that it can be tuned for capacitance so that a load capacitor is not necessary for some nodes. The part may be placed in the position where the load cap used to be for a simple change (if the cap was an 0805 part, for example).

Radio-Frequency Interference (RFI)

Most Class 2 applications must comply with radiated emissions criteria. The corresponding GM test is GM9114P.

The most important aspect of reducing Class 2 emissions is to waveshape the transceiver output adequately. The slew rate is not as important as rounding the corners.

ECU Level Testing

This section gives some insight into testing Class 2 at the ECU level.

The building block (BB) is the IC(s) that contains the Class 2 digital and analog circuitry plus the necessary peripheral components. The power supplies and grounds are included. Important BB-level tests that should be done are briefly described in the following subsections.

- Arbitration

  Correct arbitration with other Class 2 devices is done at the following:

  a. ±2% clock mismatch between nodes
  b. ±2 V ground mismatch between nodes
  c. 2 nodes and 32 nodes (lumped or distributed equivalent load)

- No Disruption of System Functions

  No disruption or corruption of Class 2 transmit or receive under the following conditions:

  a. System loss of power (and repower)
  b. System loss of ground (and reapplication of ground)
  c. System hardware or software reset
  d. Normal system activities (clocks, timers, drivers, etc.)

- No Damage from Certain Transients

  No damage to Class 2 components under the following conditions:

  a. Short any pin to power
  b. Short any pin to ground

- Symbol Widths

  Measure symbol widths with 2 nodes and with 32 nodes (lumped or distributed equivalent load):

  a. Digital widths (output of, and input to, digital section)
  b. Actual bus symbol widths

- Rise/Fall Times

  Measure bus waveform rise and fall times between 10 and 90% of the waveform height with 2 nodes and with 32 nodes (lumped or distributed equivalent load). Also measure from 1.5 to 6.25 V (V_{ol} max to V_{oh} min).

- Delays

  Measure time delays (see Fig. 9.12):

  a. Delay from transmit active to bus active
  b. Delay from transmit active to receive active
  c. Delay from bus active to receive active

  Table 9.1 gives ac parametric numbers typically required.

- System

  The Class 2 system is the "product" that contains the Class 2 building block. This product may be a radio, seat controller, and so forth. Typical systems level tests are the GM9100P series of component EMC tests. Most of these have special addenda specifically for Class 2 network testing because of the possible range of bus loading.

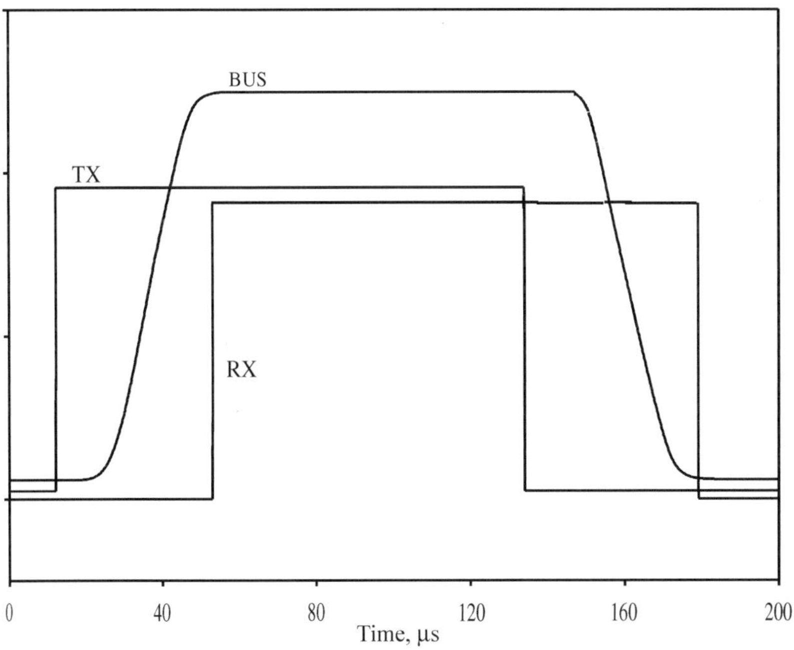

Fig. 9.12 Illustration of the delay from TX to RX.

TABLE 9.1
TYPICAL VALUES FOR DELAY FROM TX TO RX

Parameter	Definition	Conditions	Min	Typ	Max	Units
T_{osc}	Local clock tolerance	All (temp, aging, load, etc.)	–2.0	0	2.0	µs
T_t	Bus rise/fall time (measured from bus V_{ol} to V_{oh} and V_{oh} to V_{ol})	Min and max bus loading	13	—	18	µs
T_{d1}	Time delay 1 (from CL2TX V_{ih} to bus V_{oh}	Min and max bus loading	13	16	18	µs
T_{d2}	Time delay 2 (from CL2TX Vih to CL2RX Voh	Min and max bus loading	14	20	25	µs
T_{d3}	Time delay 3 (from bus V_{ih} to CL2RX V_{oh}	Min and max bus loading	1	4	7	µs

References

[1] Upender, G. "Analyzing the Real-Time Characteristics of Class C Communications in CAN Through Discrete Event Simulations. " SAE Paper No. 940133, Society of Automotive Engineers, Warrendale, PA, 1994.

[2] Demcko, R. "Multilayer Varistors in Automobile MUX Bus Applications." SAE Paper No. 981106, Society of Automotive Engineers, Warrendale, PA, 1998.

Chapter 10

The Integrated Circuit (IC) Level

Introduction

This chapter covers some examples that show how serial data link functionality is implemented at the discrete or integrated circuit (IC) level. Several mechanizations illustrating the major partitioning options are presented.

Figure 10.1 shows the level of implementation that this chapter explores. IC level components are shaded and include microcontrollers that have built-in digital/analog, serial data modules (cells) and standalone digital/analog ICs.

Partitioning

There is a wide range of choices for hardware partitioning when mechanizing a protocol handler [1]. One example of this is to place the digital portion (and possibly the transceiver portion) onboard the microcontroller, as shown in Fig. 10.2. This is usually the most cost-effective solution relative to the cost of hardware and software in interfacing the CPU to the bus. The cost of the chip will depend on what else is on it, of course. Increased processing steps to accommodate the transceiver will no doubt raise the price (if the transceiver can be integrated at all). The software should be simpler because the circuitry is onboard with the CPU (no external peripheral data buses), and the overall reliability is enhanced by a decreased chip count. This is, however, the most inflexible partitioning because of the limited amount of existing chip configurations. One is at the mercy of what is commercially available.

A standalone method of hardware partitioning, on the other hand, provides extreme flexibility in picking a CPU and other functionality of the

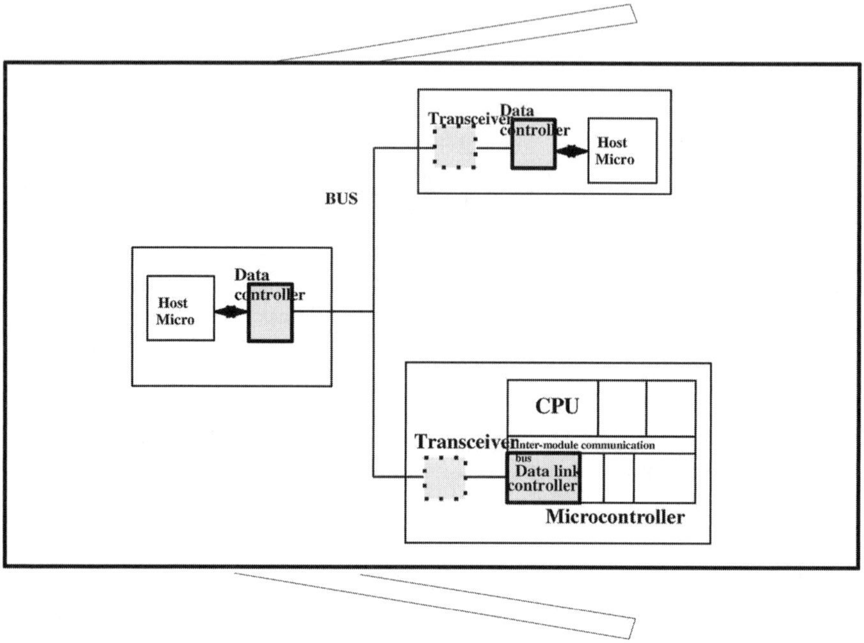

Fig. 10.1 IC level.

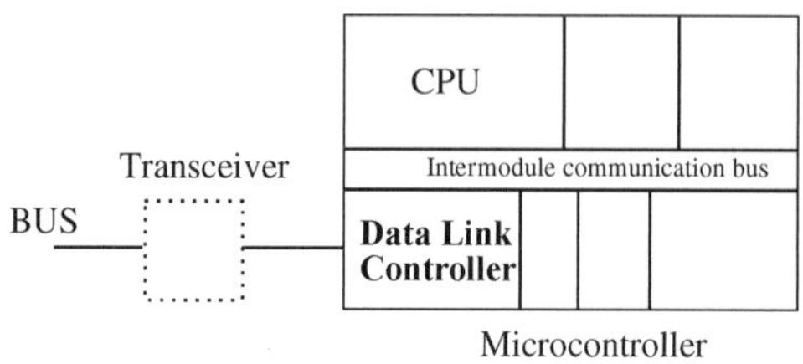

Fig. 10.2 Block diagram of up-integrated protocol handler.

microcontroller. It does have the disadvantages of increased chip count (Fig. 10.3) and more complicated software needs.

Standalone partitioning has another drawback: decreased reliability. More ICs mean more pins and more solder joints that can break loose. There are the obvious difficulties in placing, routing, and providing enough real estate for the parts.

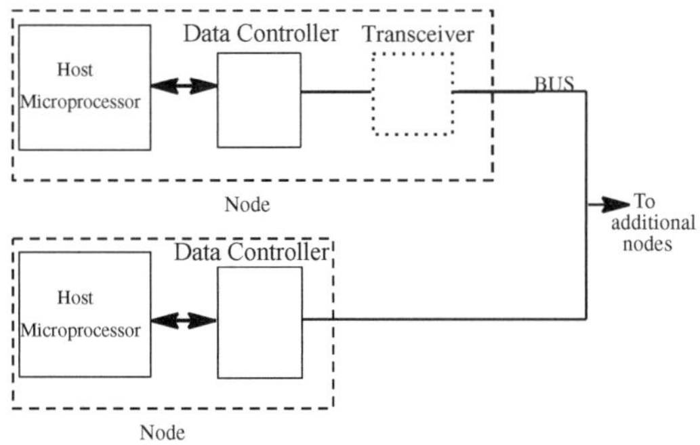

Fig. 10.3 Block diagram of a standalone protocol handler.

A somewhat extreme level of up-integration has surfaced occasionally in recent years. This involves the concept of placing the entire protocol handler within the box connector. Many fiber-optics implementations use this scheme, as shown in Fig. 10.4. It makes sense in this case especially because the LED and PIN diode should be placed close to the ends of the optical fibers. Integrating the rest of the analog and digital circuitry into the connector is not too difficult.

Connector integration has the most promise in the area of "smart sensors and actuators." In this type of multiplexing, some nodes do not contain a

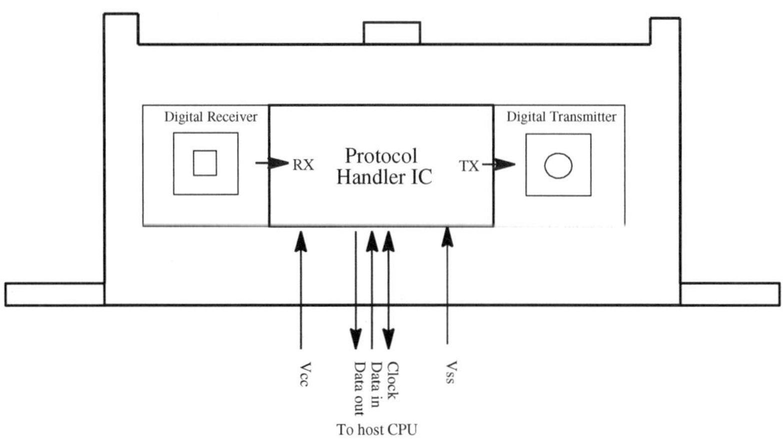

Fig. 10.4 Block diagram of a protocol handler integrated into a connector.

microprocessor (or maybe not even a clock or power supply), and probably there is no circuit board. The only place for the serial data interface is in the connector.

Compared to the other partitioning choices, the connector method is obviously very inflexible.

As mentioned while discussing levels of hardware support for bus data (i.e., message, byte, bit), hardware can be traded for software and vice versa. Historically, the lion's share of serial data interfacing duty has been placed on custom dedicated hardware. But now because software is free (only a hardware engineer's joke), recent designs have reduced the amount of silicon area needed in exchange for an increased software burden.

It is possible to implement a protocol handler entirely in software. An example of this was a project by the Rollins Group of Companies and Automotive Diagnostic Equipment Corporation in 1993 [2]. A Motorola 68332 time processor unit (TPU) was used for J1850 transmit and receive functions. The TPU is well suited for a timing-intensive protocol like J1850 but takes up most of its capabilities. It is unlikely to be able to do anything else.

Figure 10.5 indicates that more software is necessary to provide for the higher layers of the OSI model [3]. Unfortunately, this is also true for the amount of hardware required. The usual delineation is to have dedicated hardware perform the required functions in layers 1 and 2 and have software do all the rest. That may not sound fair, but most protocol specifications are heavy on layers 1 and 2. The higher layers are usually application dependent, anyway.

General Digital Trade-Offs

Table 10.1 shows some of the advantages and disadvantages of various buffer designs in multiplex ICs. Most of these attributes are debatable and not clear because they depend on what else is on the chip. Many people agree that byte level is the current norm and baseline [4].

Clocking the digital circuitry is another important issue. It is desirable to use an inexpensive oscillator such as a ceramic resonator instead of a crystal. Unfortunately, some early analyses indicated that very tight tolerance sources

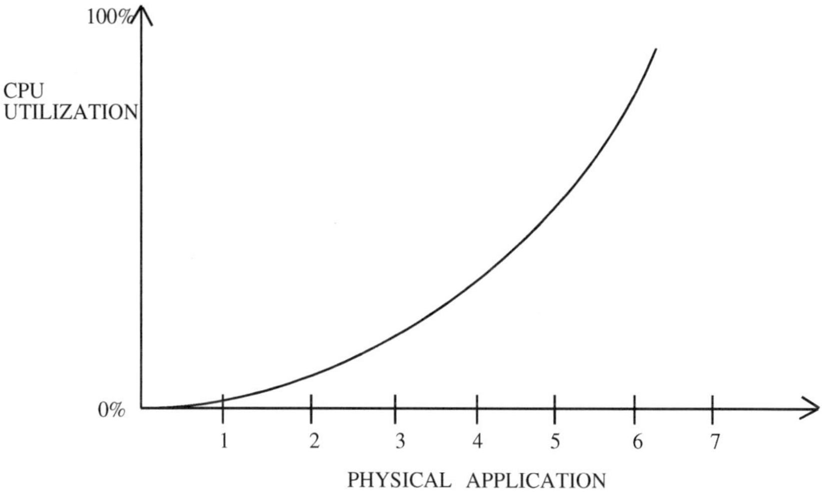

Fig. 10.5 Software load per OSI layer.

TABLE 10.1
BUFFER TRADE-OFFS IN DIGITAL IC DESIGN

	Symbol	Byte	Message
Advantages	• Cheap • Small H/W size	• Average price • Average size	• Little code
Disadvantages	• Lots of code	• Average code	• Large H/W size • Expensive
Hardware examples	• Harris 'Sendec' on '05	• Motorola BDLC on '08 • Harris JBLIC on HIP-7010 • Intel on '196LB • TI on PRISM	• DLCS/DLCP • Motorola MDLC on '05 • Motorola DLCM on '332

were going to be necessary for some protocols. For example, a paper [5] by Dais and Chapman of Bosch projected that high-speed (i.e., 1 Mb/s) CAN would require a ±0.38% clock.

General Analog Trade-Offs

Table 10.2 shows a comparison of various digital versus analog partitioning options. Most of these arguments center on the IC vendor's ability to integrate power-handling circuitry with low-power logic [4, 6].

TABLE 10.2
PARTITIONING TRADE-OFFS IN ANALOG IC DESIGN

	μ + dlc + idr	μ + idr	μ
Advantages	• Very flexible • Easiest to design/build	• Sort of flexible • Average cost • Average size • Easier to fab μ	• Cheapest (maybe) • Smallest (maybe)
Disadvantages	• Many chips • Most expensive	• Extra chip	• Very inflexible • Hard to fab

μ + idr: microcontroller including serial data digital circuitry + standalone transceiver
μ + dlc: microprocessor + standalone serial data digital/analog IC
μ + dlc + idr: microprocessor + standalone serial data digital IC + standalone transceiver

Message Filtering

Many protocols lend themselves to message filtering, and others do not. Class C protocols, such as CAN, typically provide message-level masking. Most J1850 chips, on the other hand, do not. Reference [7] provides several possible mechanisms for J1850 message filtering. Note, however, that message filtering is actually an implementation issue, not one of the protocol itself.

Digital CAN Implementations

Early on, CAN designs at the digital IC level were categorized by the speed of the bus and amount of data buffering they provided [8]. Fig. 10.6 shows roughly how this was broken down.

The Intel 82527 is an important hardware implementation of the CAN protocol [9, 10]. The 82527 is all-digital and is used as a standalone (peripheral) interface. This chip, designed with input from Bosch, replaced the 82526.

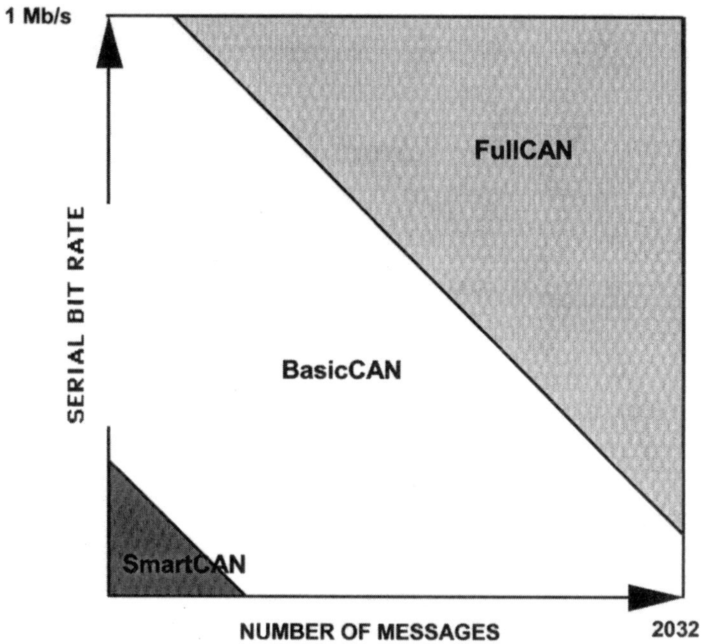

Fig. 10.6 CAN implementation levels.

The first silicon was made in late 1992. The major advantage of this IC is that it had an extensive CPU interface capability. Also, unlike the '526, the '527 allowed all 2048 different message headers, whereas the '526 allowed only 2032.

The 82527 could interface to the following:

- 12 MHz 8096
- 16 MHz 8051
- 11 MHz 8048
- 8 MHz 6801/6803
- 5 MHz 6805
- 8.4 MHz 68HC11
- Advanced 80C196
- 4.2 MHz Motorola HC11 family
- Multiplexed address/data bus
- Nonmultiplexed address/data bus

- 16.667 MHz Motorola 68332
- Asynchronous (DSACK) address/data bus
- Serial peripheral interface (SPI) bus

The 82527 required a separate transceiver, but little else in the way of peripheral components. A basic mechanization is shown in Fig. 10.7, and a view of the memory map is given in Table 10.3.

The memory map of the 82527 is set up so that the chip select and address lines are logically placed somewhere in the host CPU's addressing space. This makes the contents of the 82527's "dual port RAM" directly accessible to the programmer (with some constraints).

Space for 15 message "communication objects," or buffers, takes up much of the memory map. The predecessor to the 82527, the Intel 82526, did not use a fixed message buffer scheme. As you can see in Fig. 10.8, space for the communication objects began at offset 06h and must finish by 40h. At initialization time, the programmer would set up the buffer space as large as needed, depending on the number of transmitted/received message expected and the number of data bytes in each message.

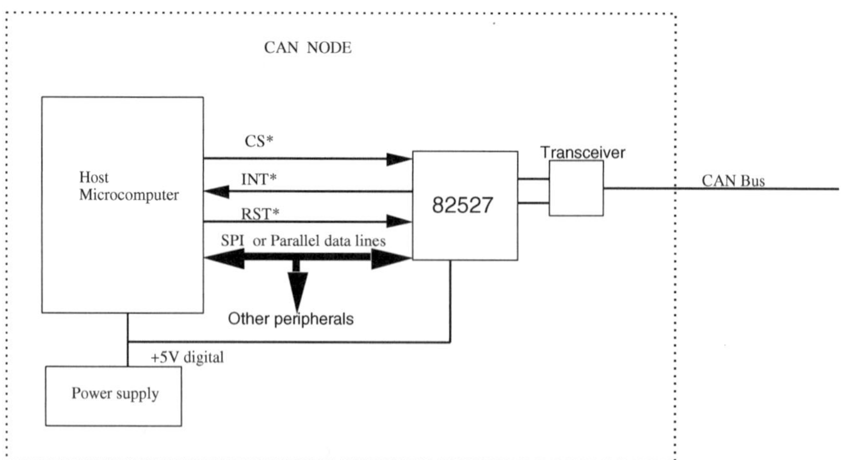

Fig. 10.7 Intel 82527 mechanization.

TABLE 10.3
THE 82527 MEMORY MAP

Address	Contents
00h	Control Register
01h	Status Register
02h	CPU Interface Register
03h	RESERVED
04h	High-speed Read (high)
05h	High-speed Read (low)
06h	Global Mask Short (high)
07h	Global Mask Short (low)
08h	Global Mask Long (high)
09h	Global Mask Long
0Ah	Global Mask Long
0Bh	Global Mask Long (low)
0Ch	Mask Last Msg (high)
0Dh	Mask Last Msg
0Eh	Mask Last Msg
0Fh	Mask Last Msg (low)
10h - 1Eh	MESSAGE #1
1Fh	Clockout Register
20h - 2Eh	MESSAGE #2
2Fh	Bus Configuration Register
30h - 3Eh	MESSAGE #3
3Fh	Bit Timing Register 0
40h - 4Eh	MESSAGE #4
4Fh	Bit Timing Register 1
50h - 5Eh	MESSAGE #5
5Fh	Interrupt Register
60h - 6Eh	MESSAGE #6
6Fh	Test Register BSP0(reserved)
70h - 7Eh	MESSAGE #7
7Fh	Test Register BSP1(reserved)
80h - 8Eh	MESSAGE #8
8Fh	Test Register BSP2(reserved)
90h - 9Eh	MESSAGE #9
9Fh	P1 Configuration
A0h - AEh	MESSAGE #10
AFh	P2 Configuration
B0h - BEh	MESSAGE #11
BFh	P1 In
C0 - CEh	MESSAGE #12
CFh	P2 In
D0h - DEh	MESSAGE #13
DFh	P1 Out
E0h - EEh	MESSAGE #14
EFh	P2 Out
F0h - FEh	MESSAGE #15
FFh	SPI Reset Address

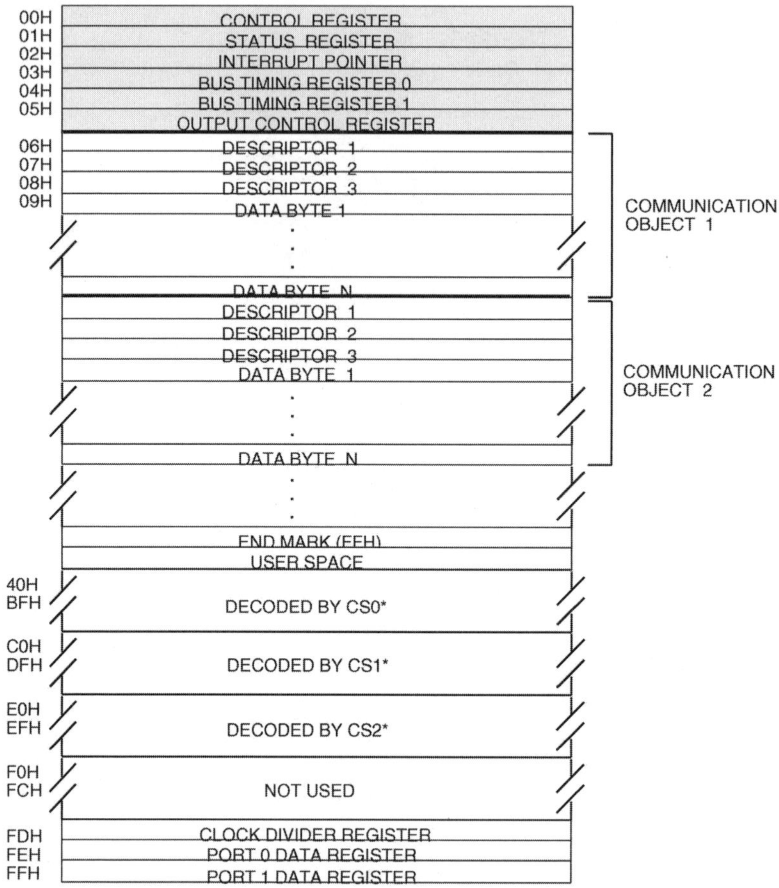

Fig. 10.8 The 82526 memory map.

Digital IC Mechanizations

Nodes that use the Motorola HC05X4 [11] use the mechanization shown in Fig. 10.9.

This is typical for up-integrated serial data interfaces. Almost all that is necessary is to hook up the transmit line and receive line. Any peripheral components are usually around the transceiver.

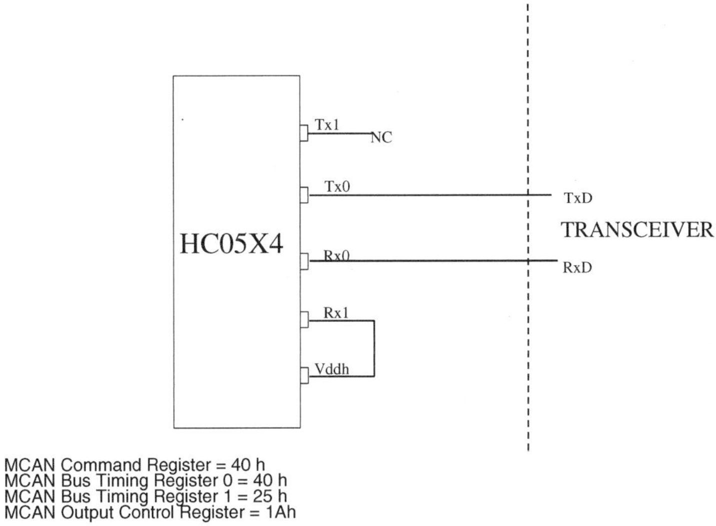

Fig. 10.9 The HC05X4 mechanization.

CAN Analog Transceivers

CAN requires some sort of physical interface in addition to the digital logic part (i.e., the 82527). The basic CAN protocol 2.0 does not specify a particular physical layer. In fact, everything from RS485 to transformer-coupling to fiber optics has been used for the physical layer of CAN designs.

Before the advent of IC transceivers, the CAN driver/receiver analog circuit was fashioned from discrete components. Probably the most important of them is the standard Bosch representation shown in Fig. 10.10. This circuit had several minor variations but was used in some production applications in the early 1990s.

This circuit became the model for the ISO 11898 physical layer [12]. It is essentially the discrete equivalent of the typical ISO 11898 IC transceiver available today, such as the Philips 82C250.

The output is depicted electrically in Fig. 10.11. Note that with ISO 11898, the two CAN lines are resistively terminated to each other. The chapter on Class C Protocols contains more information on CAN termination strategies.

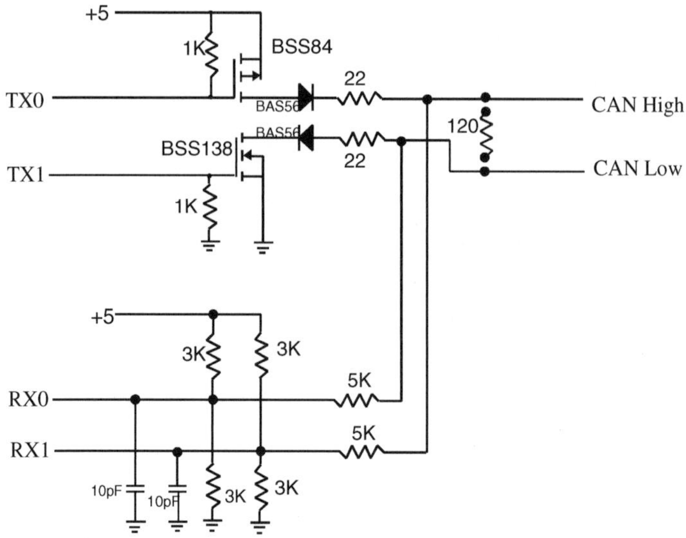

Fig. 10.10 Discrete CAN high-speed transceiver.

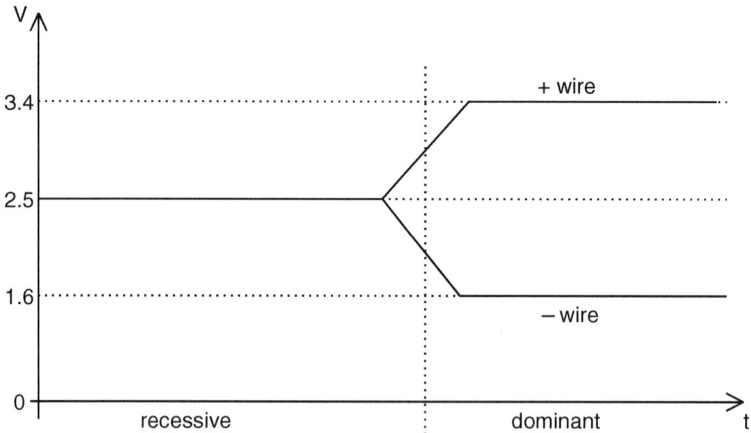

Fig. 10.11 High-speed CAN waveform.

Figure 10.12 is another type of CAN physical interface. Philips first proposed the fundamental design in 1990 [13]. It is commonly referred to as the low-speed CAN interface, or ISO 11519-2 [14]. This circuit is incompatible with the high-speed version. An electrical representation is shown in Fig. 10.13.

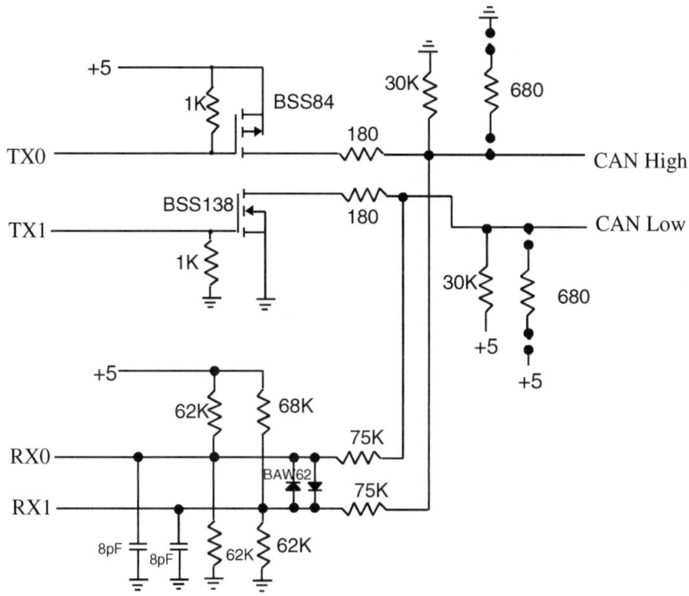

Fig. 10.12 Discrete low-speed CAN transceiver circuit.

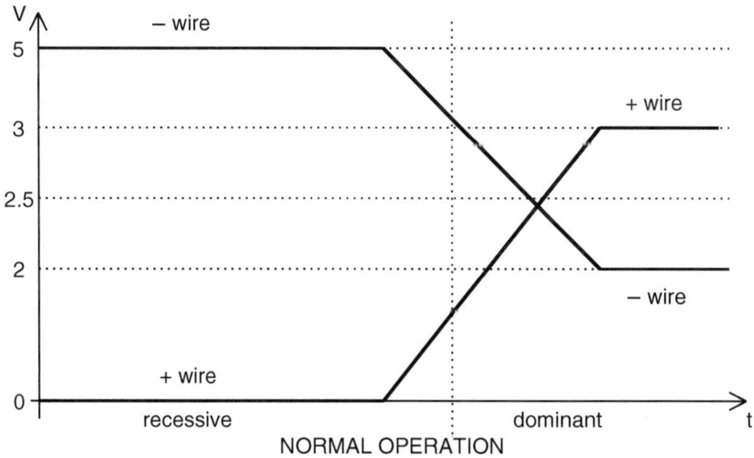

Fig. 10.13 Low-speed CAN transceiver waveform.

A low-speed CAN transceiver, such as ISO 11519, can operate better with a missing or shorted wire than the ISO 11898 version in which the CAN bus lines are terminated to one another. An example of this is illustrated in Fig. 10.14.

Analog (Transceiver) IC Mechanizations

Figure 10.15 shows a basic mechanization for, probably, the most commonly used transceiver for CAN applications—the Philips 82C250 [15]. This part provides an ISO 11898 compatible interface.

Sometimes a common mode choke is required to suppress common mode radiated frequencies [16, 17] (see Fig. 10.16).

Figure 10.17 shows one possible schematic representation of the common mode choke. The most commonly used part is the Epcos B82790-50513-N201 [18]. Other chokes are becoming available all the time. One big drawback of the chokes has been the size and cost. The Epcos part

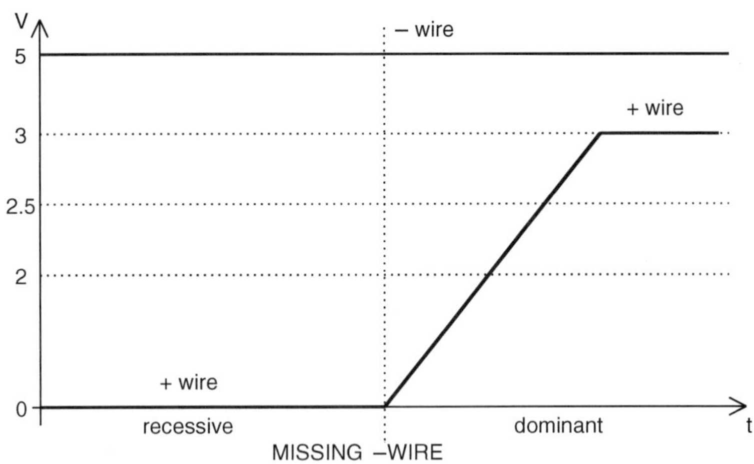

Fig. 10.14 Low-speed CAN transceiver operation with error.

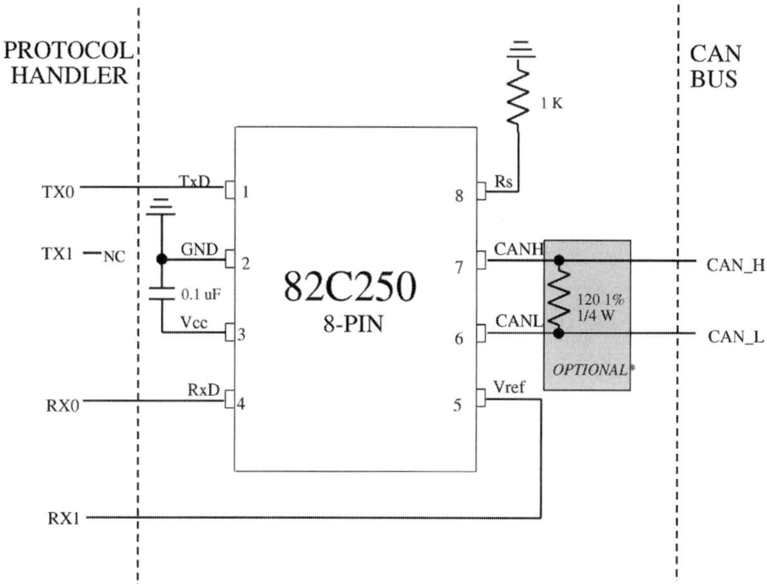

* Not all nodes contain the termination resistor.

Fig. 10.15 The 82C250 transceiver mechanization.

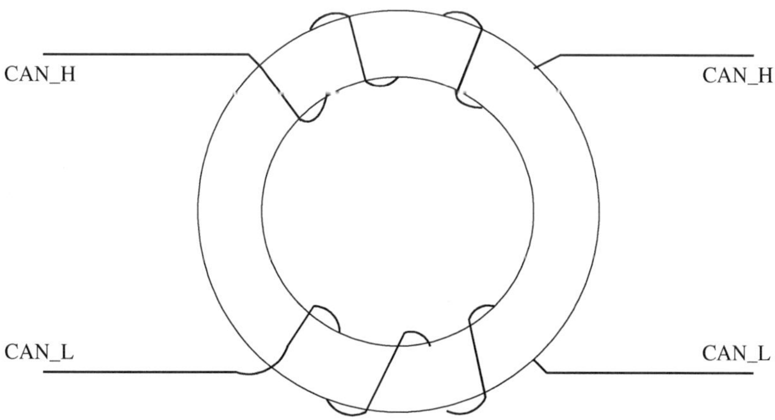

Fig. 10.16 Physical representation of common mode choke.

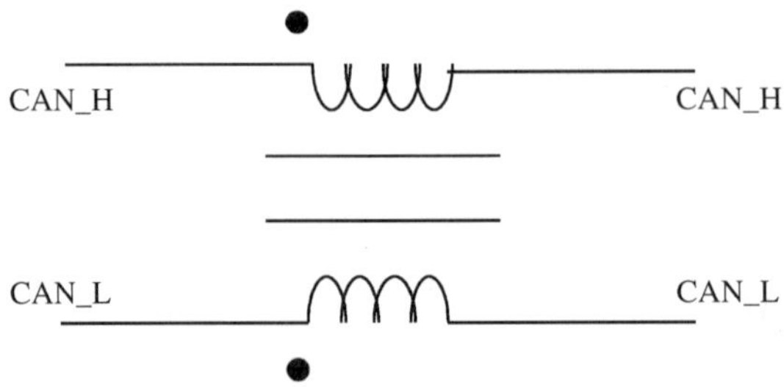

Fig. 10.17 Schematic representation of common mode choke.

referenced here is approximately a cube, 1/4 inch on a side, priced about $0.40 in large quantities.

To get a feeling for the number of commercial ICs and up-integrated cells available to implement CAN, reference Table 10.4.

Table 10.5 compares some major features of the high-speed (ISO 11898) CAN transceiver ICs available.

IC Level Testing

IC testing is done by the manufacturer per the part drawing parameters. Devices are usually not 100% tested, but a certain random percentage is checked exactly against the IC specification.

The IC level is the first line of defense in designing for good RFI, radiated susceptibility, and ESD. If there is good performance at the IC level, there will be good performance at the systems, or ECU, level. Unfortunately, after that, all bets are off. It is difficult to get good correlation between ECU level and vehicle level testing. This will be discussed more in the next chapter.

TABLE 10.4
CAN IC AND MODULE AVAILABILITY

Standalone Name	Company	Vers.	Object Storage Capacity	Acceptance Filtering Capability	Remarks
CC750	Bosch	2.0B	14 Tx/Rx buffers + 1 double Rx buffer	2 global masks + 1 mask for Rx double buffer	function compatible with Intel 82527, interface to CPU by SPI only, SOIC16-W package
CC770	Bosch	2.0B	14 Tx/Rx buffers + 1 double Rx buffer	2 global masks + 1 mask for Rx double buffer	pin and function compatible with Intel 82527, interface is 8/16-bit multiplexed/non-multiplexed
SAK 82C900	Infineon	2.0B	32 Tx/Rx buffers	32 full-bit masks	2 independent CAN nodes, 32 independent message objects, hardware gateway support, hardware FIFO support, parallel, SPI and standalone mode
MCP25020	Microchip	2.0B	3 Tx buffers + 2 Rx buffers	2 full-bit Acceptance Filters 1 Acceptance Masks	I/O extender without microcontroller
MCP25025	Microchip	2.0B	3 Tx buffers + 2 Rx buffers	2 full-bit Acceptance Filters 1 Acceptance Masks	I/O extender without microcontroller
MCP25050	Microchip	2.0B	3 Tx buffers + 2 Rx buffers	2 full-bit Acceptance Filters 1 Acceptance Masks	I/O extender without microcontroller
MCP25055	Microchip	2.0B	3 Tx buffers + 2 Rx buffers	2 full-bit Acceptance Filters 1 Acceptance Masks	I/O extender without microcontroller
MCP2510	Microchip	2.0B	3 Tx buffers + 2 Rx buffers	6 full-bit Acceptance Filters 2 Acceptance Masks	SPI
SJA1000	Philips	2.0B	1 Tx buffer + 64 Rx FIFO	1 global 32-bit mask or 2 global 16-bit masks	replaces 82C200
TC190C580	Toshiba	2.0B	15 Tx/Rx buffers + 1 Rx buffer	15 full-bit masks + 1 global mask	

TABLE 10.4 (cont.)

8-Bit Name	Company	CAN V.	Object Storage Capacity	Acceptance Filtering Capability	RAM	Package	Remarks
T83C51CC01	Atmel	2.0A/2.0B	15 Tx/Rx buffers	15 full-bit masks	1 KiB	TQFP44, PLCC44, CA-BGA64	2kB Boot Memory, A/D, UART, TTCAN level 1 support, Autobaud, 40 MHz
T89C51CC01	Atmel	2.0A/2.0B	15 Tx/Rx buffers	15 full-bit masks	1,2 KiB	TQFP44, PLCC44, CA-BGA64	2kB Boot Flash, A/D, UART, TTCAN level 1 support, Autobaud, 40 MHz
T89C51CC02	Atmel	2.0A/2.0B	5 Tx/Rx buffers	5 full-bit masks	512 B	TQFP32, PLCC28, SOIC28	2KB Boot Flash, A/D, UART, TTCAN level 1 support, Autobaud, 40 MHz
T89C51CC03	Atmel	2.0A/2.0B	15 Tx/Rx buffers	15 full-bit masks	2,2 KiB	TQFP44, PLCC44, CA-BGA64	2kB Boot Flash, A/D, UART, TTCAN level 1 support, Autobaud, 40 MHz
C8051F040	Cygnal	2.0B	32 Tx/Rx buffers	32 full-bit masks	4 KiB	TQFP100	25 MIPS, 12bit ADC, 12bit DAC, 5 16-bit Timer, 2 UART, SPI, SMBus/I2C, 64 Port-I/O
C8051F041	Cygnal	2.0B	32 Tx/Rx buffers	32 full-bit masks	4 KiB	TQFP64	25 MIPS, 12bit ADC, 12bit DAC, 5 16-bit Timer, 2 UART, SPI, SMBus/I2C, 32 Port-I/O
C8051F042	Cygnal	2.0B	32 Tx/Rx buffers	32 full-bit masks	4 KiB	TQFP100	25 MIPS, 10bit ADC, 12bit DAC, 5 16-bit Timer, 2 UART, SPI, SMBus/I2C, 64 Port-I/O
C8051F043	Cygnal	2.0B	32 Tx/Rx buffers	32 full-bit masks	4 KiB	TQFP64	25 MIPS, 10bit ADC, 12bit DAC, 5 16-bit Timer, 2 UART, SPI, SMBus/I2C, 32 Port-I/O
C8051F060	Cygnal	2.0B	32 Tx/Rx buffers	32 full-bit masks	4 KiB	TQFP100	25 MIPS, 2 16bit ADCs, 12bit DAC, 5 16-bit Timer, 2 UART, SPI, SMBus/I2C, 59 Port-I/O
C8051F061	Cygnal	2.0B	32 Tx/Rx buffers	32 full-bit masks	4 KiB	TQFP64	25 MIPS, 2 16bit ADCs, 12bit DAC, 5 16-bit Timer, 2 UART, SPI, SMBus/I2C, 24 Port-I/O
C8051F062	Cygnal	2.0B	32 Tx/Rx buffers	32 full-bit masks	4 KiB	TQFP100	25 MIPS, 2 16bit ADCs, 12bit DAC, 5 16-bit Timer, 2 UART, SPI, SMBus/I2C, 59 Port-I/O
C8051F063	Cygnal	2.0B	32 Tx/Rx buffers	32 full-bit masks	4 KiB	TQFP64	25 MIPS, 2 16bit ADCs, 12bit DAC, 5 16-bit Timer, 2 UART, SPI, SMBus/I2C, 24 Port-I/O
DS80C390	Dallas	2 x 2.0B	14 Tx/Rx buffers + 1 double Rx buffer	14 Tx/Rx buffers + 1 double Rx buffer			40 MHz clock rate, 16/32-bit math coprocessor
DS80C400	Dallas	2.0B	14 Tx/Rx buffers + 1 double Rx buffer	14 Tx/Rx buffers + 1 double Rx buffer			75 MHz clock rate, 16/32-bit math coprocessor, Ethernet MAC
505CA-4EM	Infineon	2.0B	14 Tx/Rx buffers + 1 double Rx buffer	15 full-bit masks + 1 global mask		MQFP44	ASC, 4 x PWM, 10bit ADC
505CA-4RM	Infineon	2.0B	14 Tx/Rx buffers + 1 double Rx buffer	15 full-bit masks + 1 global mask		MQFP44	ASC, 4 x PWM, 10bit ADC
C505CA-LM	Infineon	2.0B	14 Tx/Rx buffers + 1 double Rx buffer	15 full-bit masks + 1 global mask		MQFP44	ASC, 4 x PWM, 10bit ADC
C515C-8EM	Infineon	2.0B	14 Tx/Rx buffers + 1 double Rx buffer	15 full-bit masks + 1 global mask		MQFP80	SSC, ASC, 4 x PWM
C515C-8RLM	Infineon	2.0B	14 Tx/Rx buffers + 1 double Rx buffer	15 full-bit masks + 1 global mask		MQFP80	SSC, ASC, 4 x PWM
C515C-LM	Infineon	2.0B	14 Tx/Rx buffers + 1 double Rx buffer	15 full-bit masks + 1 global mask		MQFP80	SSC, ASC, 4 x PWM
PIC18C658	Microchip	2.0B	3 Tx buffers + 2 Rx buffers	6 full-bit Acceptance Filters 2 Acceptance Masks		TQFP, CERQUAD, PLCC	ADC, AUSART, I2C, SPI, 4 + WD Timers, 2 PWM, 2 Comperators
PIC18C858	Microchip	2.0B	3 Tx buffers + 2 Rx buffers	6 full-bit Acceptance Filters 2 Acceptance Masks		TQFP, CERQUAD, PLCC	ADC, AUSART, I2C, SPI, 4 + WD Timers, 2 PWM, 2 Comperators
PIC18F248	Microchip	2.0B	3 Tx buffers + 2 Rx buffers	6 full-bit Acceptance Filters 2 Acceptance Masks		SDIP28L, SOIC28L	ADC, USART, MI2C, SPI, 4 + WD Timers, 1 PWM
PIC18F258	Microchip	2.0B	3 Tx buffers + 2 Rx buffers	6 full-bit Acceptance Filters 2 Acceptance Masks		SDIP28L, SOIC28L	ADC, USART, MI2C, SPI, 4 + WD Timers, 1 PWM
PIC18F448	Microchip	2.0B	3 Tx buffers + 2 Rx buffers	6 full-bit Acceptance Filters 2 Acceptance Masks		DIP, TQFP, PLCC	ADC, USART, MI2C, SPI, 4 + WD Timers, 5 PWM
PIC18F458	Microchip	2.0B	3 Tx buffers + 2 Rx buffers	6 full-bit Acceptance Filters 2 Acceptance Masks		DIP, TQFP, PLCC	ADC, USART, MI2C, SPI, 4 + WD Timers, 5 PWM
PIC18F6580	Microchip	2.0B	3 Tx buffers + 2 Rx buffers	6 full-bit Acceptance Filters 2 Acceptance Masks		TQFP, PLCC	ADC, AUSART, MI2C, SPI, 4 + WD Timers, 5 PWM
PIC18F6680	Microchip	2.0B	3 Tx buffers + 2 Rx buffers	6 full-bit Acceptance Filters 2 Acceptance Masks		TQFP, PLCC	ADC, AUSART, MI2C, SPI, 4 + WD Timers, 5 PWM
PIC18F8580	Microchip	2.0B	3 Tx buffers + 2 Rx buffers	6 full-bit Acceptance Filters 2 Acceptance Masks		TQFP, PLCC	ADC, AUSART, MI2C, SPI, 4 + WD Timers, 5 PWM
PIC18F8680	Microchip	2.0B	3 Tx buffers + 2 Rx buffers	6 full-bit Acceptance Filters 2 Acceptance Masks		TQFP, PLCC	ADC, AUSART, MI2C, SPI, 4 + WD Timers, 5 PWM
M37630E4FP	Mitsubishi	2.0B	1 Tx buffer + 2 Rx buffers	1 full-bit mask + 1 global mask	512 B		
M37630M4T-xxxFP	Mitsubishi	2.0B	1 Tx buffer + 2 Rx buffers	1 full-bit mask + 1 global mask	512 B		
M37630M4V-xxxFP	Mitsubishi	2.0B	1 Tx buffer + 2 Rx buffers	1 full-bit mask + 1 global mask	512 B		up to 125°C
M37632EFFP	Mitsubishi	2.0B	1 Tx buffer + 2 Rx buffers	1 full-bit mask + 1 global mask	2 KiB		
M37632MCT-xxxFP	Mitsubishi	2.0B	1 Tx buffer + 2 Rx buffers	1 full-bit mask + 1 global mask	2 KiB		
M37632MFT-xxxFP	Mitsubishi	2.0B	1 Tx buffer + 2 Rx buffers	1 full-bit mask + 1 global mask	2 KiB		
68HC(7)05X32	Motorola	2.0A	1 Tx buffer + 2 Rx buffers	1 global 8-bit mask		64-pin	
68HC(7)05X4	Motorola	2.0A	1 Tx buffer + 2 Rx buffers	1 global 8-bit mask		28-pin	
68HC05X16	Motorola	2.0A	1 Tx buffer + 2 Rx buffers	1 global 8-bit mask		64-pin	
68HC08AZ32A	Motorola	2.0B	3 Tx buffers + 2 Rx buffers	1 global 32-bit or 2 global 16-bit or 4 global 8-bit masks		64-pin	
68HC08AZ60	Motorola	2.0B	3 Tx buffers + 2 Rx buffers	1 global 32-bit or 2 global 16-bit or 4 global 8-bit masks		64-pin	
68HC908AZ60A	Motorola	2.0B	3 Tx buffers + 2 Rx buffers	1 global 32-bit or 2 global 16-bit or 4 global 8-bit masks		64-pin	
COP684BC	NatSem	2.0Bp	1 Tx 2-byte buffer + 2 Rx 2-byte buffer	1 global 8-bit mask			8-byte frames can be handles up to 125 kbit/s
COP688/89EB	NatSem	2.0Bp	1 Tx 2-byte buffer + 2 Rx 2-byte buffer	1 global 8-bit mask			8-byte frames can be handles up to 125 kbit/s
COP87/L88/89EB	NatSem	2.0Bp	1 Tx 2-byte buffer + 2 Rx 2-byte buffer	1 global 8-bit mask			OTP Family;8-byte frames can be handles up to 125 kbit/s
COP87L84BC	NatSem	2.0Bp	1 Tx 2-byte buffer + 2 Rx 2-byte buffer	1 global 8-bit mask			OTP Family; 8-byte frames can be handles up to 125 kbit/s
COP888/89EB	NatSem	2.0Bp	1 Tx 2-byte buffer + 2 Rx 2-byte buffer	1 global 8-bit mask			8-byte frames can be handles up to 125 kbit/s
COP8884BC	NatSem	2.0Bp	1 Tx 2-byte buffer + 2 Rx 2-byte buffer	1 global 8-bit mask			8-byte frames can be handles up to 125 kbit/s
CANASSP1 - µPD780948	NEC	2.0B	2 Tx buffers + 16 Rx buffers	16 full-bit masks + 2 global 29-bit masks	2 KiB	QFP100	6 timers, LCD, CAN Time System, DCAN
CANASSP1 - µPD78F0948	NEC	2.0B	2 Tx buffers + 16 Rx buffers	16 full-bit masks + 2 global 29-bit masks	2 KiB	QFP100	6 timers, LCD, CAN Time System, DCAN
CANASSP2 - µPD 78F0818	NEC	2.0B	2 Tx buffers + 16 Rx buffers	16 full-bit masks + 2 global 29-bit masks	3 KiB	QFP64	6 timers, 2xCSI, UART, CAN Time System, DCAN
CANASSP2 - µPD780814/6	NEC	2.0B	2 Tx buffers + 16 Rx buffers	16 full-bit masks + 2 global 29-bit masks	1,5 KiB	QFP64	6 timers, 2xCSI, UART, CAN Time System, DCAN
CANASSP3 - µPD 78F0828	NEC	2.0B	2 Tx buffers + 16 Rx buffers	16 full-bit masks + 2 global 29-bit masks	3 KiB	QFP80	6 timers, CSI, UART, LCD, Stepper motor driver, CAN Time System, DCAN
CANASSP3 - µPD780824/6/8	NEC	2.0B	2 Tx buffers + 16 Rx buffers	16 full-bit masks + 2 global 29-bit masks	1,5, 3 KiB	QFP80	6 timers, CSI, UART, LCD, Stepper motor driver, CAN Time System, DCAN
E60 - µPD780701/3Y	NEC	2.0B	2 Tx buffers + 16 Rx buffers	16 full-bit masks + 2 global 29-bit masks	3 KiB	QFP80	16 x 8-bit A/D, 7 timers, DCAN, 2xCSI, UART, I2C, 3 x 8-bit PWM-outputs
E60 - µPD78F0701/3Y	NEC	2.0B	2 Tx buffers + 16 Rx buffers	16 full-bit masks + 2 global 29-bit masks	3 KiB	QFP80	16 x 8-bit A/D, 7 timers, DCAN, 2xCSI, UART, I2C, 3 x 8-bit PWM-outputs
MiniCAN - µPD789850	NEC	2.0B	2 Tx buffers + 14 Rx buffers	14 full-bit masks + 2 global 29-bit masks	0,7 KiB	SSOP30	3 timers, UART or CSI, CAN Time System, DCAN
MiniCAN - µPD78F9850	NEC	2.0B	2 Tx buffers + 14 Rx buffers	14 full-bit masks + 2 global 29-bit masks	0,7 KiB	SSOP30	3 timers, UART or CSI, CAN Time System, DCAN

TABLE 10.4 (cont.)

8-Bit Name	Company	CAN V.	Object Storage Capacity	Acceptance Filtering Capability	RAM	Package	Remarks
P89LPC99X	Philips	2.0B	1 Tx buffer + 64 Rx FIFO	1 global 32-bit mask or 2 global 16-bit masks			Higher layer protocolls supported by software
P8XC591	Philips	2.0B	1 Tx buffer + 64 Rx FIFO	1 global 32-bit mask or 2 global 16-bit masks			Higher layer protocolls supported by software
P8XC592/8	Philips	2.0A	1 Tx buffer + 2 Rx buffers	1 global 8-bit mask			
ST72F521	ST	1 x 2.0Bp	3 Tx/Rx buffers	2 global 11-bit masks		TQFP64, TQFP80	
ST72F561	ST	1 x 2.0B	2 Tx, 1 RX FIFO (3 msg)	12 ext. or 24 std or 48 8-bit filters		TQFP32, TQFP44, TQFP64	
ST92F150	ST	2 x 2.0B	3 Tx, 2 RX FIFO (3 msg each)	16 ext. or 32 std or 64 8-bit filters		TQFP64, TQFP100	E3 flash
ST92F250	ST	2 x 2.0B	3 Tx, 2 RX FIFO (3 msg each)	16 ext. or 32 std or 64 8-bit filters		TQFP100	E3 flash

TABLE 10.4 (cont.)

16-Bit Name	Company	CAN Version	Object Storage Capacity	Acceptance Filtering Capability	Package	Remarks
MB90387	Fujitsu	1 x 2.0B	8 Tx/Rx buffers	8 full-bit masks + 2 global masks	QFP48	UART, A/D
MB90423G	Fujitsu	2 x 2.0B	16 Tx/Rx buffers	16 full-bit masks + 2 global masks	QFP100	UARTx2, A/D, 4 Stepper Motor Drivers, LCD
MB90427G	Fujitsu	1 x 2.0B	16 Tx/Rx buffers	16 full-bit masks + 2 global masks	QFP100	UARTx2, A/D, 4 Stepper Motor Drivers, LCD
MB90428G	Fujitsu	1 x 2.0B	16 Tx/Rx buffers	16 full-bit masks + 2 global masks	QFP100	UARTx2, A/D, 4 Stepper Motor Drivers, LCD
MB90497G	Fujitsu	1 x 2.0B	8 Tx/Rx buffers	8 full-bit masks + 2 global masks	QFP64	UART, A/D
MB90547G	Fujitsu	1 x 2.0B	16 Tx/Rx buffers	16 full-bit masks + 2 global masks	QFP100	UARTx2, A/D, Ext Bus Interface
MB90548G	Fujitsu	1 x 2.0B	16 Tx/Rx buffers	16 full-bit masks + 2 global masks	QFP100	UARTx2, A/D, Ext Bus Interface
MB90549G	Fujitsu	1 x 2.0B	16 Tx/Rx buffers	16 full-bit masks + 2 global masks	QFP100	UARTx2, A/D, Ext Bus Interface
MB90591G	Fujitsu	2 x 2.0B	16 Tx/Rx buffers	16 full-bit masks + 2 global masks	QFP100	UARTx3, A/D, 4 Stepper Motor Drivers
MB90594G	Fujitsu	2 x 2.0B	16 Tx/Rx buffers	16 full-bit masks + 2 global masks	QFP100	UARTx3, A/D, 4 Stepper Motor Drivers
MB90598G	Fujitsu	1 x 2.0B	16 Tx/Rx buffers	16 full-bit masks + 2 global masks	QFP100	UARTx2, A/D, 4 Stepper Motor Drivers
MB90F347	Fujitsu	1 x 2.0B	16 Tx/Rx buffers	16 full-bit masks + 2 global masks	QFP100	UART (LIN) x4, A/D 16ch, 8ch PPG, Ext. Bus Interface
MB90F387	Fujitsu	1 x 2.0B	8 Tx/Rx buffers	8 full-bit masks + 2 global masks	QFP48	UART, A/D
MB90F394H	Fujitsu	2 x 2.0B	16 Tx/Rx buffers	16 full-bit masks + 2 global masks	QFP120	UARTx3, A/D, 6 Stepper Motor Drivers
MB90F423G	Fujitsu	2 x 2.0B	16 Tx/Rx buffers	16 full-bit masks + 2 global masks	QFP100	UARTx2, A/D, 4 Stepper Motor Drivers, LCD
MB90F428G	Fujitsu	1 x 2.0B	16 Tx/Rx buffers	16 full-bit masks + 2 global masks	QFP100	UARTx2, A/D, 4 Stepper Motor Drivers, LCD
MB90F443G	Fujitsu	3 x 2.0B	16 Tx/Rx buffers	16 full-bit masks + 2 global masks	QFP100	UARTx2, A/D, Ext Bus Interface
MB90F497G	Fujitsu	1 x 2.0B	8 Tx/Rx buffers	8 full-bit masks + 2 global masks	QFP64	UART, A/D
MB90F543G	Fujitsu	2 x 2.0B	16 Tx/Rx buffers	16 full-bit masks + 2 global masks	QFP100	UARTx2, A/D, Ext Bus Interface
MB90F546G	Fujitsu	1 x 2.0B	16 Tx/Rx buffers	16 full-bit masks + 2 global masks	QFP100	UARTx2, A/D, Ext Bus Interface
MB90F548G	Fujitsu	1 x 2.0B	16 Tx/Rx buffers	16 full-bit masks + 2 global masks	QFP100	UARTx2, A/D, Ext Bus Interface
MB90F591G	Fujitsu	2 x 2.0B	16 Tx/Rx buffers	16 full-bit masks + 2 global masks	QFP100	UARTx3, A/D, 4 Stepper Motor Drivers
MB90F594G	Fujitsu	2 x 2.0B	16 Tx/Rx buffers	16 full-bit masks + 2 global masks	QFP100	UARTx3, A/D, 4 Stepper Motor Drivers
MB90F598G	Fujitsu	1 x 2.0B	16 Tx/Rx buffers	16 full-bit masks + 2 global masks	QFP100	UARTx2, A/D, 4 Stepper Motor Drivers
H8S/2622M	Hitachi	1 x 2.0B	15 Tx/Rx buffers + 1 Rx buffers	16 full-bit masks + 1 global mask	QFP100	20MHz, SCI x 3, A/D x 16, TPU, Dual Power Supply
H8S/2623F	Hitachi	1 x 2.0B	15 Tx/Rx buffers + 1 Rx buffers	16 full-bit masks + 1 global mask	QFP100	20MHz, SCI x 3, A/D x 16, TPU, Dual Power Supply
H8S/2623M	Hitachi	1 x 2.0B	15 Tx/Rx buffers + 1 Rx buffers	16 full-bit masks + 1 global mask	QFP100	20MHz, SCI x 3, A/D x 16, TPU, Dual Power Supply
H8S/2626F	Hitachi	1 x 2.0B	15 Tx/Rx buffers + 1 Rx buffers	16 full-bit masks + 1 global mask	QFP100	20MHz, SCI x 3, A/D x 16, TPU, Sub-clock, Dual Power Supply
H8S/2635M	Hitachi	2 x 2.0B	15 Tx/Rx buffers + 1 Rx buffers	16 full-bit masks + 1 global mask	QFP128	20MHz, SCI x 3, A/D x 12, TPU, 10bit PWM + Stepper Motor Drivers, Sub-clock, Single Power Supply
H8S/2636F	Hitachi	2 x 2.0B	15 Tx/Rx buffers + 1 Rx buffers	16 full-bit masks + 1 global mask	QFP128	20MHz, SCI x 3, A/D x 12, TPU, 10bit PWM + Stepper Motor Drivers, Sub-clock, Single Power Supply
H8S/2636M	Hitachi	2 x 2.0B	15 Tx/Rx buffers + 1 Rx buffers	16 full-bit masks + 1 global mask	QFP128	20MHz, SCI x 3, A/D x 12, TPU, 10bit PWM + Stepper Motor Drivers, Sub-clock, Single Power Supply
H8S/2638F	Hitachi	2 x 2.0B	15 Tx/Rx buffers + 1 Rx buffers	16 full-bit masks + 1 global mask	QFP128	20MHz, SCI x 3, IIC, A/D x 12, TPU, 10bit PWM + Stepper Motor Drivers, Sub-clock, Single Power Supply
H8S/2639F	Hitachi	2 x 2.0B	15 Tx/Rx buffers + 1 Rx buffers	16 full-bit masks + 1 global mask	QFP128	20MHz, SCI x 3, IIC, A/D x 12, TPU, 10bit PWM + Stepper Motor Drivers, Internal sub-clock, Single Power Supply
H8S/2645M	Hitachi	1 x 2.0B	15 Tx/Rx buffers + 1 Rx buffers	16 full-bit masks + 1 global mask	QFP144	20MHz, SCI x 2, LCD(24seg x 4com), A/D x 12, TPU, 10bit PWM + Stepper Motor Drivers, Sub-clock, Single Power Supply
H8S/2646F	Hitachi	1 x 2.0B	15 Tx/Rx buffers + 1 Rx buffers	16 full-bit masks + 1 global mask	QFP144	20MHz, SCI x 2, LCD(24seg x 4com), A/D x 12, TPU, 10bit PWM + Stepper Motor Drivers, Sub-clock, Single Power Supply
H8S/2646M	Hitachi	1 x 2.0B	15 Tx/Rx buffers + 1 Rx buffers	16 full-bit masks + 1 global mask	QFP144	20MHz, SCI x 2, LCD(24seg x 4com), A/D x 12, TPU, 10bit PWM + Stepper Motor Drivers, Sub-clock, Single Power Supply
H8S/2648F	Hitachi	1 x 2.0B	15 Tx/Rx buffers + 1 Rx buffers	16 full-bit masks + 1 global mask	QFP144	20MHz, SCI x 3, LCD(40seg x 4com), A/D x 12, TPU, 10bit PWM + Stepper Motor Drivers, Sub-clock, Single Power Supply
H8S/2648M	Hitachi	1 x 2.0B	15 Tx/Rx buffers + 1 Rx buffers	16 full-bit masks + 1 global mask	QFP144	20MHz, SCI x 3, LCD(40seg x 4com), A/D x 12, TPU, 10bit PWM + Stepper Motor Drivers, Sub-clock, Single Power Supply
HMS30C7202	Hynix	2 x 2.0B	32 Tx/Rx buffers	32 full-bit masks	MQFP256, PGA256	72 MIPS, 3 32-bit Timer, 1 64-bit Timer, 4 UART, SPI, I2S
C161CS-32RF	Infineon	2.0B	2 x (14 Tx/Rx buffers + 1 double Rx buffer)	2 x (15 full-bit masks + 1 global mask)	TQFP128	I2C, 2 x ASC, SSC, RTC, 10bit ADC
C161JC-32RF	Infineon	2.0B	14 Tx/Rx buffers + 1 double Rx buffer	15 full-bit masks + 1 global mask	TQFP128	I2C, 2 x ASC, SSC, RTC, 10bit ADC, J1850 VPW
C164CI-8EM	Infineon	2.0B	14 Tx/Rx buffers + 1 double Rx buffer	15 full-bit masks + 1 global mask	MQFP80	ASC, SSC, 6 x PWM, RTC, 10bit ADC
C164CI-8RM	Infineon	2.0B	14 Tx/Rx buffers + 1 double Rx buffer	15 full-bit masks + 1 global mask	MQFP80	ASC, SSC, 6 x PWM, RTC, 10bit ADC
C164CI-LM	Infineon	2.0B	14 Tx/Rx buffers + 1 double Rx buffer	15 full-bit masks + 1 global mask	MQFP80	ASC, SSC, 6 x PWM, RTC, 10bit ADC
C164CL-8RM	Infineon	2.0B	14 Tx/Rx buffers + 1 double Rx buffer	15 full-bit masks + 1 global mask	MQFP80	ASC, SSC, 6 x PWM (reduced), RTC, 10bit ADC
C164CM-4E25F	Infineon	2.0B	14 Tx/Rx buffers + 1 double Rx buffer	15 full-bit masks + 1 global mask	TQFP64	ASC, SSC, RTC, 6 x PWM, 10bit ADC, 25 MHz
C164CM-4EF	Infineon	2.0B	14 Tx/Rx buffers + 1 double Rx buffer	15 full-bit masks + 1 global mask	TQFP64	ASC, SSC, RTC, 6 x PWM, 10bit ADC, 20 MHz
C164CM-4R25F	Infineon	2.0B	14 Tx/Rx buffers + 1 double Rx buffer	15 full-bit masks + 1 global mask	TQFP64	ASC, SSC, RTC, 6 x PWM, 10bit ADC, 25 MHz
C164CM-4RF	Infineon	2.0B	14 Tx/Rx buffers + 1 double Rx buffer	15 full-bit masks + 1 global mask	TQFP64	ASC, SSC, RTC, 6 x PWM, 10bit ADC, 20 MHz
C167CR-16RM	Infineon	2.0B	14 Tx/Rx buffers + 1 double Rx buffer	15 full-bit masks + 1 global mask	MQFP144	ASC, SSC, 4xPWM, RTC, 10bit ADC
C167CR-4RM	Infineon	2.0B	14 Tx/Rx buffers + 1 double Rx buffer	15 full-bit masks + 1 global mask	MQFP144	ASC, SSC, 4xPWM, RTC, 10bit ADC
C167CR-LM	Infineon	2.0B	14 Tx/Rx buffers + 1 double Rx buffer	15 full-bit masks + 1 global mask	MQFP144	ASC, SSC, 4xPWM, RTC, 10bit ADC
C167CS-4RM	Infineon	2.0B	2 x (14 Tx/Rx buffers + 1 double Rx buffer)	2 x (15 full-bit masks + 1 global mask)	MQFP144	ASC, SSC, 4xPWM, RTC, 10bit ADC
C167CS-LM	Infineon	2.0B	2 x (14 Tx/Rx buffers + 1 double Rx buffer)	2 x (15 full-bit masks + 1 global mask)	MQFP144	ASC, SSC, 4xPWM, RTC, 10bit ADC
XC161CJ-16FF	Infineon	2.0B	32 Tx/Rx buffers	32 full-bit masks	TQFP144	I2C, 2 x ASC, 2 x SSC, RTC, 12bit ADC, J1850 VPW, DSP Support,
XC164CS-16F20F	Infineon	2.0B	32 Tx/Rx buffers	32 full-bit masks	TQFP100	6 x PWM, 1 x ASC, 2 x SSC, RTC, 10bit ADC, DSP Support, OCDS,
XC164CS-16F40F	Infineon	2.0B	32 Tx/Rx buffers	32 full-bit masks	TQFP100	6 x PWM, 1 x ASC, 2 x SSC, RTC, 10bit ADC, DSP Support, OCDS,
XC164CS-16R20F	Infineon	2.0B	32 Tx/Rx buffers	32 full-bit masks	TQFP100	6 x PWM, 1 x ASC, 2 x SSC, RTC, 10bit ADC, DSP Support, OCDS,

TABLE 10.4 (cont.)

16-Bit Name	Company	CAN Version	Object Storage Capacity	Acceptance Filtering Capability	Package	Remarks
XC164CS-16R40F	Infineon	2.0B	32 Tx/Rx buffers	32 full-bit masks	TQFP100	6 x PWM, 1 x ASC, 2 x SSC, RTC, 10bit ADC, DSP Support, OCDS,
XC164CS-8F20F	Infineon	2.0B	32 Tx/Rx buffers	32 full-bit masks	TQFP100	6 x PWM, 1 x ASC, 2 x SSC, RTC, 10bit ADC, DSP Support, OCDS,
XC164CS-8F40F	Infineon	2.0B	32 Tx/Rx buffers	32 full-bit masks	TQFP100	6 x PWM, 1 x ASC, 2 x SSC, RTC, 10bit ADC, DSP Support, OCDS,
XC164CS-8R20F	Infineon	2.0B	32 Tx/Rx buffers	32 full-bit masks	TQFP100	6 x PWM, 1 x ASC, 2 x SSC, RTC, 10bit ADC, DSP Support, OCDS,
XC164CS-8R40F	Infineon	2.0B	32 Tx/Rx buffers	32 full-bit masks	TQFP100	6 x PWM, 1 x ASC, 2 x SSC, RTC, 10bit ADC, DSP Support, OCDS,
XC167CI-16F	Infineon	2.0B	32 Tx/Rx buffers	32 full-bit masks	TQFP144	6 x PWM, 2 x ASC, 2 x SSC, RTC, 10bit ADC, DSP Support, OCDS
M301N2F8TFP	Mitsubishi	2.0B	16 Tx/Rx buffers	16 full-bit masks + 3 global masks		
M301N2F8TFP	Mitsubishi	2.0B	16 Tx/Rx buffers	16 full-bit masks + 3 global masks		
M301N2M4T-xxxFP	Mitsubishi	2.0B	16 Tx/Rx buffers	16 full-bit masks + 3 global masks		
M301N2M4T-xxxFP	Mitsubishi	2.0B	16 Tx/Rx buffers	16 full-bit masks + 3 global masks		
M306N4FCTFP	Mitsubishi	2 x 2.0B	2 x 16 Tx/Rx buffers	2 x (16 full-bit masks + 3 global masks)		3 phase motor controllers
M306N4FCVFP	Mitsubishi	2 x 2.0B	2 x 16 Tx/Rx buffers	2 x (16 full-bit masks + 3 global masks)		3 phase motor controllers, up to 125°C
M306N4FGTFP	Mitsubishi	2 x 2.0B	2 x 16 Tx/Rx buffers	2 x (16 full-bit masks + 3 global masks)		3 phase motor controllers
M306N4FGVFP	Mitsubishi	2 x 2.0B	2 x 16 Tx/Rx buffers	2 x (16 full-bit masks + 3 global masks)		3 phase motor controllers, up to 125°C
M306N4MCT-xxxFF	Mitsubishi	2 x 2.0B	2 x 16 Tx/Rx buffers	2 x (16 full-bit masks + 3 global masks)		3 phase motor controllers
M306N4MCV-xxxFF	Mitsubishi	2 x 2.0B	2 x 16 Tx/Rx buffers	2 x (16 full-bit masks + 3 global masks)		3 phase motor controllers, up to 125°C
M306N4MGT-xxxFF	Mitsubishi	2 x 2.0B	2 x 16 Tx/Rx buffers	2 x (16 full-bit masks + 3 global masks)		3 phase motor controllers
M306N4MGV-xxxFF	Mitsubishi	2 x 2.0B	2 x 16 Tx/Rx buffers	2 x (16 full-bit masks + 3 global masks)		3 phase motor controllers, up to 125°C
M306N5FCTFP	Mitsubishi	2.0B	16 Tx/Rx buffers	16 full-bit masks + 3 global masks		3 phase motor controllers
M306N5FCVFP	Mitsubishi	2.0B	16 Tx/Rx buffers	16 full-bit masks + 3 global masks		3 phase motor controllers, up to 125°C
M306N5MCT-xxxFF	Mitsubishi	2.0B	16 Tx/Rx buffers	16 full-bit masks + 3 global masks		3 phase motor controllers
M306N5MCV-xxxFF	Mitsubishi	2.0B	16 Tx/Rx buffers	16 full-bit masks + 3 global masks		3 phase motor controllers, up to 125°C
M306NAFGTFP	Mitsubishi	2 x 2.0B	2 x 16 Tx/Rx buffers	2 x (16 full-bit masks + 3 global masks)		3 phase motor controllers
M306NAMCT-xxxFF	Mitsubishi	2 x 2.0B	2 x 16 Tx/Rx buffers	2 x (16 full-bit masks + 3 global masks)		3 phase motor controllers
M306NAMCV-xxxFF	Mitsubishi	2 x 2.0B	2 x 16 Tx/Rx buffers	2 x (16 full-bit masks + 3 global masks)		3 phase motor controllers, up to 125°C
M306NAMGT-xxxFF	Mitsubishi	2 x 2.0B	2 x 16 Tx/Rx buffers	2 x (16 full-bit masks + 3 global masks)		3 phase motor controllers
M306NBFCTFP	Mitsubishi	2.0B	16 Tx/Rx buffers	16 full-bit masks + 3 global masks		3 phase motor controllers
M306NBFCTFP	Mitsubishi	2.0B	16 Tx/Rx buffers	16 full-bit masks + 3 global masks		3 phase motor controllers
M306NBMCT-xxxFF	Mitsubishi	2.0B	16 Tx/Rx buffers	16 full-bit masks + 3 global masks		3 phase motor controllers
M306NBMCT-xxxFF	Mitsubishi	2.0B	16 Tx/Rx buffers	16 full-bit masks + 3 global masks		3 phase motor controllers
M306NBMCV-xxxFF	Mitsubishi	2.0B	16 Tx/Rx buffers	16 full-bit masks + 3 global masks		3 phase motor controllers, up to 125°C
M306NBMCV-xxxFF	Mitsubishi	2.0B	16 Tx/Rx buffers	16 full-bit masks + 3 global masks		3 phase motor controllers, up to 125°C
68HC912BC32	Motorola	2.0B	3 Tx buffers + 2 Rx buffers	1 global 32-bit or 2 global 16-bit or 4 global 8-bit masks	80-pin	
68HC912D60	Motorola	2.0B	3 Tx buffers + 2 Rx buffers	1 global 32-bit or 2 global 16-bit or 4 global 8-bit masks	80-pin, 112-pin	
68HC912DG128	Motorola	2 x 2.0B	3 Tx buffers + 2 Rx buffers	2 global 32-bit or 4 global 16-bit or 8 global 8-bit masks	112-pin	
68HC912DT128A	Motorola	3 x 2.0B	3 Tx buffers + 2 Rx buffers	2 global 32-bit or 4 global 16-bit or 8 global 8-bit masks	112-pin	
MC9S12DP256	Motorola	5 x 2.0B	3 Tx buffers + 4 Rx buffers	2 global 32-bit or 4 global 16-bit or 8 global 8-bit masks	80-pin, 112-pin	
CR16BTC	NatSem	2.0B	15 Tx/Rx buffers + 1 Rx buffer	14 full-bit masks + 1 global mask		Incl. Bluetooth, USA, I2C, USART
CR16HCS9VJE	NatSem	2.0B	15 Tx/Rx buffers + 1 Rx buffer	14 full-bit masks + 1 global mask		Flash memory, 24 MHz, I2C, 8-bit A/D, up to 2 USART's, Extended temperature range
CR16HCT9VJE	NatSem	2.0B	15 Tx/Rx buffers + 1 Rx buffer	14 full-bit masks + 1 global mask		Flash memory, 96 kB, 24 MHz, I2C, 8-bit A/D, up to 2 USART's, Extended temperature range
CR16MAS5VJE	NatSem	2.0B	15 Tx/Rx buffers + 1 Rx buffer	14 full-bit masks + 1 global mask		ROM based, I2C, 8-bit A/D, up to 2 USART´s, Extended temperature range
CR16MAS9VJE	NatSem	2.0B	15 Tx/Rx buffers + 1 Rx buffer	14 full-bit masks + 1 global mask		Flash memory, Extended temperature range
CR16MBR5VJE	NatSem	2.0B	15 Tx/Rx buffers + 1 Rx buffer	14 full-bit masks + 1 global mask		ROM based, I2C, 8-bit A/D, up to 2 USART´s, Extended temperature range
CR16MCS5VJE	NatSem	2.0B	15 Tx/Rx buffers + 1 Rx buffer	14 full-bit masks + 1 global mask		ROM based, I2C, 8-bit A/D, up to 2 USART´s, Extended temperature range
CR16MCS9VJE	NatSem	2.0B	15 Tx/Rx buffers + 1 Rx buffer	14 full-bit masks + 1 global mask		Flash memory, I2C, 8-bit A/D, up to 2 USART´s, Extended temperature range
XA-C3	Philips	2.0B	32 Tx/Rx buffers	32 full-bit masks + global masks		CANopen, DeviceNet and OSEK transport layer hardwired
ST10F168	ST	1 x 2.0B	14 Tx/Rx buffers + 1 double Rx buffer	15 full-bit masks + 1 global mask	TQFP144	
ST10F269	ST	2 x 2.0B	14 Tx/Rx buffers + 1 double Rx buffer	15 full-bit masks + 1 global mask	TQFP144,PQFP144	
ST92F150	ST	2 x 2.0B	3 Tx, 2 RX FIFO (3 msg each)	16 ext. or 32 std or 64 8-bit filters	TQFP64,TQFP100	E3 flash
ST92F250	ST	2 x 2.0B	3 Tx, 2 RX FIFO (3 msg each)	16 ext. or 32 std or 64 8-bit filters	TQFP100	E3 flash
TMP86CH87U	Toshiba	2.0B	7 Tx/Rx buffers + 1 Rx buffer	7 full-bit masks + 1 global mask		
TMP86PH87U	Toshiba	2.0B	7 Tx/Rx buffers + 1 Rx buffer	7 full-bit masks + 1 global mask		
TMP91CM80F	Toshiba	2.0B	15 Tx/Rx buffers + 1 Rx buffer	15 full-bit masks + 1 global mask		
TMP91PP80F	Toshiba	2.0B	15 Tx/Rx buffers + 1 Rx buffer	15 full-bit masks + 1 global mask		
TMP95CS54F	Toshiba	2.0B	15 Tx/Rx buffers + 1 Rx buffer	15 full-bit masks + 1 global mask		
TMP95CU54F	Toshiba	2.0B	15 Tx/Rx buffers + 1 Rx buffer	15 full-bit masks + 1 global mask		
TMP95FW54F	Toshiba	2.0B	15 Tx/Rx buffers + 1 Rx buffer	15 full-bit masks + 1 global mask		
TMP95PS54F	Toshiba	2.0B	15 Tx/Rx buffers + 1 Rx buffer	15 full-bit masks + 1 global mask		

TABLE 10.4 (cont.)

32-Bit Name	Company	CAN V.	Object Storage Cap.	Acceptance Filtering Capability	Package	Remarks
MB91F362GA	Fujitsu	3 x 2.0B	16 Tx/Rx buffers	16 full-bit masks + 2 global masks	LQFP208	UARTx3, SIOx2, I2C, A/D, D/A, 4 Stepper Motor Drivers, Timer, Ext. Bus, 64MHz
MB91F364G	Fujitsu	1 x 2.0B	16 Tx/Rx buffers	16 full-bit masks + 2 global masks	LQFP120	UARTx1, LIN-UARTx2, SIO, I2C, A/D, D/A, Timer, Subclock, 64MHz, LQFP120
MB91F365GB	Fujitsu	2 x 2.0B	16 Tx/Rx buffers	16 full-bit masks + 2 global masks	LQFP120	UARTx2, SIOx2, I2C, A/D, D/A, 4 Stepper Motor Drivers, Timer, 32MHz
MB91F366GB	Fujitsu	2 x 2.0B	16 Tx/Rx buffers	16 full-bit masks + 2 global masks	LQFP120	UARTx2, SIOx2, I2C, A/D, 4 Stepper Motor Drivers, Subclock, Timer, 32MHz
MB91F367GB	Fujitsu	2 x 2.0B	16 Tx/Rx buffers	16 full-bit masks + 2 global masks	LQFP120	UARTx2, SIOx2, I2C, A/D, Timer, 32MHz
MB91F368GB	Fujitsu	2 x 2.0B	16 Tx/Rx buffers	16 full-bit masks + 2 global masks	LQFP120	UARTx2, SIOx2, I2C, A/D, Subclock, Timer, 32MHz
MB91F369GA	Fujitsu	2 x 2.0B	16 Tx/Rx buffers	16 full-bit masks + 2 global masks	LQFP160	UARTx1, SIOx2, I2C, A/D, Timer, Ext. Bus, 64MHz
MB91F376G	Fujitsu	2 x 2.0B	16 Tx/Rx buffers	16 full-bit masks + 2 global masks	LQFP120	UARTx2, SIOx2, I2C, A/D, 4 Stepper Motor Drivers, Subclock, Timer, 32MHz
SH7047F	Hitachi	1 x 2.0B	31 Tx/Rx buffers + 1 Rx buffer	32 full-bit masks + 32 global masks	QFP100	40MHz, SCI x 3, A/D x 16, MMT, AUD, CAN with TT function, Single Power Supply
SH7049M	Hitachi	1 x 2.0B	31 Tx/Rx buffers + 1 Rx buffer	32 full-bit masks + 32 global masks	QFP100	40MHz, SCI x 3, A/D x 16, MMT, AUD, CAN with TT function, Single Power Supply
SH7052F	Hitachi	1 x 2.0B	15 Tx/Rx buffers + 1 Rx buffer	16 full-bit masks + 1 global mask	QFP208	40MHz, SCI x 5, A/D x 16, ATU-II, AUD, Dual Power Supply
SH7053F	Hitachi	1 x 2.0B	15 Tx/Rx buffers + 1 Rx buffer	16 full-bit masks + 1 global mask	QFP208	40MHz, SCI x 5, A/D x 16, ATU-II, AUD, Dual Power Supply
SH7054F	Hitachi	1 x 2.0B	15 Tx/Rx buffers + 1 Rx buffer	16 full-bit masks + 1 global mask	QFP208	40MHz, SCI x 5, A/D x 16, ATU-II, AUD, Dual Power Supply
SH7055F	Hitachi	2 x 2.0B	15 Tx/Rx buffers + 1 Rx buffer	16 full-bit masks + 1 global mask	QFP256	40MHz, SCI x 5, A/D x 32, ATU-II, AUD, Dual Power Supply
SH705xF	Hitachi	2 x 2.0B	31 Tx/Rx buffers + 1 Rx buffer	32 full-bit masks + 32 global masks	QFP256	40MHz, SCI x 5, A/D x 32, ATU-II, AUD, CAN with TT function
TC1765-L40N	Infineon	2.0B	32 Tx/Rx buffers	32 full-bit masks	LBGA260	DMA-Controller, 2 x ASC, 2 x SSC, GPTA, 2 x 8/10/12-bit ADC, J1850 VPW, DSP Support, OCDS level 1
TC1765-L40T	Infineon	2.0B	32 Tx/Rx buffers	32 full-bit masks	LBGA260	DMA-Controller, 2 x ASC, 2 x SSC, GPTA, 2 x 8/10/12-bit ADC, J1850 VPW, DSP Support, OCDS level 2
TC1775-40LE	Infineon	2.0B	32 Tx/Rx buffers	32 full-bit masks	BGA329	PCP, 2 x ASC, 2 x SSC, GPTA, GPTU, 2 x 8/10/12-bit ADC, J1850 VPW, DSP Support, OCDS level 2
TC1912	Infineon	2.0B	32 Tx/Rx buffers	32 full-bit masks	LBGA208	64KiB Cache, 3 x ASC supports IrDA, SSC, RTC, Codec, IIC, DSP Support, OCDS
TC1920	Infineon	2.0B	32 Tx/Rx buffers	32 full-bit masks	LBGA260	64KiB Cache, 3 x ASC supports IrDA, SSC, RTC, Codec, 2 x 8/10/12 bit ADC, J1850 VPW, IIC, DSP Support, OCDS
CBC3207A	Micronas	4 x 2.0B	32 Tx/Rx buffers per CAN	32 full-bit masks + 1 global mask per CAN		512k Byte Flash, 4 independent CAN-Modules, 32 time-stamps p. CAN
CDC3252G	Micronas	2 x 2.0B	32 Tx/Rx buffers per CAN	32 full-bit masks + 1 global mask per CAN		256k Byte ROM, 1 independent CAN-Modules, 32 time-stamps p. CAN
CDC3272G	Micronas	2 x 2.0B	32 Tx/Rx buffers	32 full-bit masks + 1 global mask		384k Byte ROM, 2 independent CAN-Modules, 32 time-stamps p. CAN
CEA3207A	Micronas	2 x 2.0B	32 Tx/Rx buffers per CAN	32 full-bit masks + 1 global mask per CAN		512k Byte Flash, 2 independent CAN-Modules, 32 time-stamps p. CAN
M30833FJTFP	Mitsubishi	2.0B	16 Tx/Rx buffers	16 full-bit masks + 3 global masks		512KByte Flash, 31KByte RAM, up to 125°C
M30835FJTFP	Mitsubishi	2.0B	16 Tx/Rx buffers	16 full-bit masks + 3 global masks		512KByte Flash, 31KByte RAM, up to 125°C
M30850FJFP	Mitsubishi	2 x 2.0B	2 x 16 Tx/Rx buffers	2 x (16 full-bit masks + 3 global masks)		512KByte Flash, 16KByte RAM
M30852FJGP	Mitsubishi	2 x 2.0B	2 x 16 Tx/Rx buffers	2 x (16 full-bit masks + 3 global masks)		512KByte Flash, 16KByte RAM
M30853FHFP	Mitsubishi	2 x 2.0B	2 x 16 Tx/Rx buffers	2 x (16 full-bit masks + 3 global masks)		384KByte Flash, 24KByte RAM
M30855FHGP	Mitsubishi	2 x 2.0B	2 x 16 Tx/Rx buffers	2 x (16 full-bit masks + 3 global masks)		384KByte Flash, 24KByte RAM
M32174F3VFP	Mitsubishi	2.0B	16 Tx/Rx buffers	16 full-bit masks + 3 global masks		384KByte Flash, 40KByte RAM, up to 125°C
M32174F3VWG	Mitsubishi	2.0B	16 Tx/Rx buffers	16 full-bit masks + 3 global masks		384KByte Flash, 40KByte RAM, up to 125°C
M32174F4VFP	Mitsubishi	2.0B	16 Tx/Rx buffers	16 full-bit masks + 3 global masks		512KByte Flash, 40KByte RAM, up to 125°C
M32174F4VWG	Mitsubishi	2.0B	16 Tx/Rx buffers	16 full-bit masks + 3 global masks		512KByte Flash, 40KByte RAM, up to 125°C
M32180F8TFP	Mitsubishi	2 x 2.0B	2 x 16 Tx/Rx buffers	2 x (16 full-bit masks + 3 global masks)		1024KByte Flash, 48KByte RAM
M32180F8VFP	Mitsubishi	2 x 2.0B	2 x 16 Tx/Rx buffers	2 x (16 full-bit masks + 3 global masks)		1024KByte Flash, 48KByte RAM, up to 125°C
M32182F2TFP	Mitsubishi	2 x 2.0B	2 x 16 Tx/Rx buffers	2 x (16 full-bit masks + 3 global masks)		256KByte Flash, 64KByte RAM
M32182F2VFP	Mitsubishi	2 x 2.0B	2 x 16 Tx/Rx buffers	2 x (16 full-bit masks + 3 global masks)		256KByte Flash, 64KByte RAM, up to 125°C
MC68376	Motorola	2.0B	16 Tx/Rx buffers	16 full-bit masks + 3 global masks		time-stamp
MC68377 (EXCALIBU	Motorola	2.0B	16 Tx/Rx buffers	16 full-bit masks + 3 global masks		time-stamp
MC68F375 (CERBERI	Motorola	2.0B	16 Tx/Rx buffers	16 full-bit masks + 3 global masks		time-stamp, 256K flash memory
MPC555 (BLACK OAK	Motorola	2 x 2.0B	32 Tx/Rx buffers	32 full-bit masks + 6 global masks		2 independent on-chip CAN modules, time-stamp, 448K flash memory
MPC561 (SILVER OAI	Motorola	3 x 2.0B	48 Tx/Rx buffers	48 full-bit masks + 9 global masks		3 independent on-chip CAN modules, time-stamp, 0K flash memory
MPC563 (GREEN OAI	Motorola	3 x 2.0B	48 Tx/Rx buffers	48 full-bit masks + 9 global masks		3 independent on-chip CAN modules, time-stamp, 512K flash memory
MPC565 (SPANISH O	Motorola	3 x 2.0B	48 Tx/Rx buffers	48 full-bit masks + 9 global masks		3 independent on-chip CAN modules, time-stamp, 1M flash memory
ATOMIC - μPD703123	NEC	2.0B	64 Tx/Rx buffers	64 full-bit masks + 4 global 29-bit masks□for each FCAN	QFP144	256k Mask-ROM, 10k RAM, 7 timers, LCD, CAN Time System, 3x FCAN, 2xCSI, 3xUART, 12ch x 10bit AD, 144 QFP
ATOMIC - μPD70F312	NEC	2.0B	64 Tx/Rx buffers	64 full-bit masks + 4 global 29-bit masks□for each FCAN	QFP144	256k Flash, 10k RAM, 7 timers, LCD, CAN Time System, 3x FCAN, 2xCSI, 3xUART, 12ch x 10bit AD, 144 QFP
AVALON - μPD703080	NEC	2.0B	2 Tx buffers + 16 Rx buffers	16 full-bit masks + 2 global 29-bit masks □for each DCA	QFP128	128k Mask-ROM, 6k RAM, 3x16-bit timers, 2x8-bit timers, 2xDCAN, 3xCSI, 2xUART, 8ch x 10bit AD, LCD, stepper motor driver, 128 QFP
AVALON - μPD70F30	NEC	2.0B	2 Tx buffers + 16 Rx buffers	16 full-bit masks + 2 global 29-bit masks □for each DCA	QFP128	128k Flash, 6k RAM, 3x16-bit timers, 2x8-bit timers, 2xDCAN, 3xCSI, 2xUART, 8ch x 10bit AD, LCD, stepper motor driver, 128 QFP
BODY - μPD703xxx	NEC	2.0B	2 Tx buffers + 16 Rx buffers	16 full-bit masks + 2 global 29-bit masks	QFP100	128k Mask-ROM, 6k RAM, 2x16-bit timers, 3x8-bit timers, DCAN, 2xCSI, 2xUART, 12ch x 10bit AD, 100 QFP
BODY - μPD70F3xxx	NEC	2.0B	2 Tx buffers + 16 Rx buffers	16 full-bit masks + 2 global 29-bit masks	QFP100	128k Flash, 6k RAM, 2x16-bit timers, 3x8-bit timers, DCAN, 2xCSI, 2xUART, 12ch x 10bit AD, 100 QFP

TABLE 10.4 (cont.)

32-Bit Name	Company	CAN V.	Object Storage Cap.	Acceptance Filtering Capability	Package	Remarks
CANAIR - μPD703144	NEC	2.0B	32 Tx/Rx buffers	32 full-bit masks + 4 global 29-bit masks for each FCAN	QFP100	128k Mask-ROM, 8kRAM, 4x16-bit timers; FCAN, 2xCSI, 2xUART, 14ch x 10bit AD, 100 QFP
CANAIR - μPD70F314	NEC	2.0B	32 Tx/Rx buffers	32 full-bit masks + 4 global 29-bit masks for each FCAN	QFP100	128k Flash, 8kRAM, 4x16-bit timers; FCAN, 2xCSI, 2xUART, 14ch x 10bit AD, 100 QFP
CARGATE - μPD70F3	NEC	2.0B	64 Tx/Rx buffers	64 full-bit masks + 4 global 29-bit masks for each FCAN	QFP100	128k Flash, 8k RAM, 2x16-bit timers, CAN Time System, 5x FCAN, 3xCSI, 2xUART, 4ch x 10bit AD, 100 QFP
PHOENIX - μPD70311	NEC	2.0B	32 Tx/Rx buffers	32 full-bit masks + 4 global 29-bit masks	QFP144	256k Mask-ROM, 10k RAM, 7x16-bit timers (inverter control PWM output); FCAN, 2xCSI, 3xUART, 2 x 8ch x 10bit AD, 144pin QFP
PHOENIX - μPD70F31	NEC	2.0B	32 Tx/Rx buffers	32 full-bit masks + 4 global 29-bit masks	QFP144	256k Flash, 10k RAM, 7x16-bit timers (inverter control PWM output); FCAN, 2xCSI, 3xUART, 2 x 8ch x 10bit AD, 144pin QFP
SC3 - μPD703089Y	NEC	2.0B	32 Tx/Rx buffers	32 full-bit masks + 4 global 29-bit masks for each FCAN	QFP144	512k Mask-ROM, 24k RAM, 10x16-bit timers; 2xFCAN, 6xCSI, 2xUART (selectable), 2xI2C (selectable), 12ch x 10bit AD,144 QFP
SC3 - μPD70F3089Y	NEC	2.0B	32 Tx/Rx buffers	32 full-bit masks + 4 global 29-bit masks for each FCAN	QFP144	512k Flash, 24k RAM, 10x16-bit timers; 2xFCAN, 6xCSI, 2xUART (selectable), 2xI2C (selectable), 12ch x 10bit AD,144 QFP
SF1 - μPD703075Y	NEC	2.0B	32 Tx/Rx buffers	32 full-bit masks + 4 global 29-bit masks for each FCAN	QFP100	128k Mask-ROM, 10kRAM, 8x16-bit timers; 1xFCAN, 4xCSI, 2xUART, 1 x I2C, 12ch x 10bit AD, 100 QFP
SF1 - μPD703076Y	NEC	2.0B	32 Tx/Rx buffers	32 full-bit masks + 4 global 29-bit masks for each FCAN	QFP100	256k Mask-ROM, 10kRAM, 8x16-bit timers; 2xFCAN, 4xCSI, 2xUART, 1 x I2C, 12ch x 10bit AD, 100 QFP
SF1 - μPD703078Y	NEC	2.0B	32 Tx/Rx buffers	33 full-bit masks + 4 global 29-bit masks for each FCAN	QFP100	256k Mask-ROM, 16kRAM, 8x16-bit timers; 1xFCAN, 4xCSI, 2xUART, 1 x I2C, 12ch x 10bit AD, 100 QFP
SF1 - μPD703079Y	NEC	2.0B	32 Tx/Rx buffers	32 full-bit masks + 4 global 29-bit masks for each FCAN	QFP100	256k Mask-ROM, 16kRAM, 8x16-bit timers; 2xFCAN, 4xCSI, 2xUART, 1 x I2C, 12ch x 10bit AD, 100 QFP
SF1 - μPD70F3077Y	NEC	2.0B	32 Tx/Rx buffers	32 full-bit masks + 4 global 29-bit masks for each FCAN	QFP100	128kFlash, 10kRAM, 8x16-bit timers; 2xFCAN, 4xCSI, 2xUART, 1 x I2C, 12ch x 10bit AD, 100 QFP
SF1 - μPD70F3079Y	NEC	2.0B	32 Tx/Rx buffers	32 full-bit masks + 4 global 29-bit masks for each FCAN	QFP100	256kFlash, 16kRAM, 8x16-bit timers; 2xFCAN, 4xCSI, 2xUART, 1 x I2C, 12ch x 10bit AD,100 QFP
TMP92CW53F	Toshiba	2.0B	15 Tx/Rx buffers + 1 Rx buffer	15 full-bit masks + 1 global mask		128kB Mask ROM
TMP94FY53F	Toshiba	2.0B	15 Tx/Rx buffers + 1 Rx buffer	15 full-bit masks + 1 global mask		256kB Flash ROM
TMPR3903BF	Toshiba	2.0B	30 Tx/Rx buffers + 2 Rx buffer	30 full-bit masks + 2 global mask		analog graphics ctr. for driver information / telematic systems

TABLE 10.4 (cont.)

DSP Name	Company	CAN Version	Processor	Object Storage Cap.	Acceptance Filtering Capability	RAM	Flash	Package	Remarks	Website
ADSP21992	Analog Devices	2.0B	16-bit DSP	16 Tx/Rx buffers	16 full-bit masks	96 KiB			8 x 14-bit ADC, PWM, SPI	www.analog.com/
DSP56F803	Motorola	2.0 A/B (MSCAN)	16-bit (DSP/MCU	3 Tx buffers + 2 Rx buffers	2 global 32-bit or 4 global 16-bit or 8 global 8-bit masks		76 KiB	100-pin		http://www.mot-sps.com
DSP56F805	Motorola	2.0 A/B (MSCAN)	16-bit (DSP/MCU	3 Tx buffers + 2 Rx buffers	2 global 32-bit or 4 global 16-bit or 8 global 8-bit masks		76 KiB	144-pin		http://www.mot-sps.com
DSP56F807	Motorola	2.0 A/B (MSCAN)	16-bit (DSP/MCU	3 Tx buffers + 2 Rx buffers	2 global 32-bit or 4 global 16-bit or 8 global 8-bit masks		136 KiB	160-pin		http://www.mot-sps.com
DSP56F835	Motorola	2.0 A/B (FlexCAN)	16-bit (DSP/MCU	up to 32 Tx/Rx buffers	Three Programmable Mask Registers: Global (for MBs 0-13), special for MB14 and special for MB15.		140 KiB	144-pin		http://www.mot-sps.com
DSP56F837	Motorola	2.0 A/B (FlexCAN)	16-bit (DSP/MCU	up to 32 Tx/Rx buffers	Three Programmable Mask Registers: Global (for MBs 0-13), special for MB14 and special for MB15.		280 KiB	144-pin		http://www.mot-sps.com
DSP56F838	Motorola	2.0 A/B (FlexCAN)	16-bit (DSP/MCU	up to 32 Tx/Rx buffers	Three Programmable Mask Registers: Global (for MBs 0-13), special for MB14 and special for MB15.		280 KiB	160-pin		http://www.mot-sps.com
TMS320F241	Texas Instrumen	2.0B	16-bit DSP	6 Tx/Rx buffers	2 full-bit masks + 1 global mask	1088 B	16 KiB		8 x 10-bit A/D, 8 x PWM, QEP, SCI, SPI	www.ti.com
TMS320F243	Texas Instrumen	2.0B	16-bit DSP	6 Tx/Rx buffers	2 full-bit masks + 1 global mask	1088 B	16 KiB		8 x 10-bit A/D, 8 x PWM, QEP, SCI, SPI	www.ti.com
TMS320LC2406	Texas Instrumen	2.0B	16-bit DSP	6 Tx/Rx buffers	2 full-bit masks + 1 global mask	5 KiB			16 x 10-bit A/D, 16 x PWM, QEP, SCI, SPI	www.ti.com
TMS320LF2406	Texas Instrumen	2.0B	16-bit DSP	6 Tx/Rx buffers	2 full-bit masks + 1 global mask	5 KiB	64 KiB		16 x 10-bit A/D, 16 x PWM, QEP, SCI, SPI	www.ti.com
TMS320LF2407	Texas Instrumen	2.0B	16-bit DSP	6 Tx/Rx buffers	2 full-bit masks + 1 global mask	5 KiB	64 KiB		16 x 10-bit A/D, 16 x PWM, QEP, SCI, SPI	www.ti.com

TABLE 10.4 (cont.)

IP Name	Company	CAN V.	Object Storage Capacity	Acceptance Filtering Capability	Remarks
A_CAN	Bosch	2.0B	3 Tx/Rx buffers	1 full-bit mask	Synthesizable IP module, CAN module with 3 configurable message buffers, flexible interface for 8-bit, 16-bit, or 32-bit CPU
C_CAN	Bosch	2.0B	32 Tx/Rx buffers	32 full-bit masks	Synthesizable IP module, CAN module with message RAM for storage of 32 (or more) message objects, flexible interface for 8-bit, 16-bit, or 32-bit CPU
CAN_Core	Bosch	2.0B	n.a.	n. a.	Synthesizable IP module, CAN core cell with protocol controller and shift register - included in A_CAN, C_CAN, and TTCAN
TTCAN	Bosch	2.0B & ISO11898-1,4	32 Tx/Rx buffers	32 full-bit masks	Synthesizable IP module, C_CAN module with fully integrated TTCAN protocol functions, ISO 11898-4 Level 1 & Level 2
CANaccess	Inicore Inc.	2.0B	1 Rx buffer / 1 Tx buffer	none	FPGA core solution with full trace information, for in-depth CAN analysis, with access to all internal bits
CANmodule-II	Inicore Inc.	2.0B	3 Tx msg. w priority arbitration / 4 msg. Rx buffer	each buffer w. sep. filter, covering ID, IDE, RTR, 2 first Data Bytes	for all integration technologies (FPGA & ASIC), with AMBA (APB, 8-bit, 16-bit, 32-bit) compliant interface
CANmodule-III	Inicore Inc.	2.0B	8 Tx msg. w progr. priority arbit. / 16 msg. Rx buffer	each buffer w. sep. filter, covering ID, IDE, RTR, 2 first Data Bytes	for all integration technologies (FPGA & ASIC), with AMBA (APB, 8-bit, 16-bit, 32-bit) compliant interface
CANmodule-IIr	Inicore Inc.	2.0B	3 Tx msg. w priority arbitration / 4 msg. Rx buf	each buffer w. sep. filter, covering ID, IDE, RTR, 2 first Data Bytes	same then CANmodule-II, except receive FIFO is SRAM based
iniCAN	Inicore Inc.	2.0B	Customer specific add-ons	Customer specific add-ons	for all integration technologies (FPGA & ASIC)
FCAN	sci-worx	2.0B	configurable from 8 to 32 RX/TX buffers	configurable, including Device Net Filtering	Synthesizable IP module (verilog) for ASIC and FPGA, flexible interface for 8 bit and 16 bit, features to chose: number of mailboxes, number of masks, Acceptance filtering on data field, Univeral 16 bit counter, Sleep mode, Transmit Priority by Identifier, programmable Error Warning, etc.
FCAN - Extended	sci-worx	2.0B	16 Tx/Rx buffers	16 full-bit masks, including first 2 data byte masks	Synthesizable IP module, flexible 8/16 bit interface, Universal Counter, ErrorStatus, Programmable Error Warning, Transmit Priority by Identifier, Databyte Filtering(DeviceNet), Sleep Mode
FCAN - Highend	sci-worx	2.0B	8 Tx-only + 16 Tx/Rx + 8 Rx-only buffers	32 full-bit masks, including first 2 data byte masks	Synthesizable IP module, flexible 8/16 bit interface, features like FCAN - Extended
FCAN - Highend,A	sci-worx	2.0B	8 Tx-only + 16 Tx/Rx + 8 Rx-only buffers	32 full-bit masks	Synthesizable IP module, like FCAN Highend but with AMBA AHB interface
FCAN - Small	sci-worx	2.0B	2 Tx-only + 4 Tx/Rx + 2 Rx-only buffers	1 local full-bit mask + 1 global mask	Synthesizable IP module, flexible 8/16 bit interface
FCAN - Standard	sci-worx	2.0B	4 Tx-only + 8 Tx/Rx + 4 Rx-only buffers	2 local full-bit masks + 1 global mask	Synthesizable IP module, flexible 8/16 bit interface

TABLE 10.5
CAN TRANSCEIVER COMPARISON

	PHILIPS 82C250	PHILIPS 82C251	SILICONIX Si9200DY	TI SN65LBC031	BOSCH CF150B
SPECIFICATION	9/94 Data Sheet	10/26/95 Data Sheet	9/12/94	5/92 Advance Info.	7/95 Technical Data
PACKAGE(S)	8-pin SO 8-pin DIP	8-pin SO 8-pin DIP	8-pin SO	8-pin SO	8-pin SO
TEMP RANGE	-40° C +125° C	-40° C +125° C	-40° C +85° C	-40° C +125° C	-40° C +110° C
SUPPLY (V) Normal Maximum	 4.5 5.5 -0.3 9.0	 4.5 5.5 -0.3 9.0	 4.75 5.25 -0.3 18	 4.5 5.5 -2 7	 4.5 5.5 -0.3 7
SUPPLY CURRENT (mA): Standby Recessive Driven	 0.100 0.170 14 (w/s) 18 (HS) 70	 0.200 0.275 10 (w/s) 80	 No Standby 15 25 60 75	 No Standby 15 20 70 80	 No Standby 0 20 0 80
MAX SPEED	1 Mb/s	1 Mb/s	1 Mb/s	500 Kb/s	500 Kb/s
DRIVE CURRENT CAN_H (mA)	35	40	–	50	–
MAX VOLTAGE (V) CAN_H, CAN_L	-8 18	-40 40	-3 16	-5 20	-5 36
VOLTAGES (V) CAN_H CAN_L	 2.0 4.5 0.5 3.0	 3.0 4.5 0.5 2.0	 2.0 4.5 0.5 3.0	 2.0 4.5 0.5 3.0	 1.8 4.5 0.5 3.3
COMMON MODE RANGE (V)	-2 7	-2 7	-2 +7	–	-5 +12
DIFFERENTIAL VOLTAGE (V) Passive Driven	 -0.5 +0.05 1.5 3.0	 -0.5 +0.05 1.5 3.0	 -0.5 +0.05 1.5 3	 -0.5 +0.05 1.5 3	 -0.5 +0.05 1.5 3.0
PROP DELAY(ns): Transmitter (HS) Receiver (HS)	 50 80 50 110	 50 80 50 110	 25 50 35 70	 40 140	 50 150
SHORT CIRCUIT CURRENT (mA)	160	200	150	150	150
INPUT LOAD: Resistance (Ω) Capacitance (pF)	 10 K 25 K 20	 10 K 25 K 20	 5K 50 K 50	 5 K 50 K 20	 9.5 K 35 K 25
WAVESHAPING (V/μs)	1)Hi Spd (20-50) 2)Waveshape(3-15) 3)Standby(0)	1)Hi Spd (20-50) 2)Waveshape(3-15) 3)Standby(0)	none	1)<500Kb/s(20-40) 2)<125Kb/s(4-14)	1)<500Kb/s(20-50) 2)<125Kb/s(5-20)
BUS TRANSIENTS: ESD Overvoltage	 +/- 2 KV -150 +100 V	 +/- 2.5 KV -200 +200 V	 +2 KV -60 V +60 V	 – -150 V +100 V	 +/- 2 KV -150 +100
OTHER FEATURES	Thermal Shutdown, Low Power Shutdown, Short to +18V, -8V	Thermal Shutdown, Low Power Shutdown, Short to +40V, -40V	Short to +16V, -3V	Thermal Shutdown, Short to +20V, -5V, 50 ns receive filter	Overcurrent Shutdown Short to -5V, +36V

TABLE 10.5 (cont.)

	Mietec MTC-3054	National DS36C250	ST L9615	Unitrode UC5350	Infineon
SPECIFICATION		2/96 Advance Info.	3/96 Target Spec.	3/98 Preliminary	2.01 3/30/99
PACKAGE(S)	16-pin SO	8-pin SO mini DIP	8-pin SO	8-pin SO 8-pin DIP	8-pin SO
TEMP RANGE		-40° C +125° C	-40° C +110° C	-40° C +125° C	-40° C +150° C
SUPPLY (V) Normal Maximum		 4.5 5.5 – 12	 4.5 5.5 -0.3 7	 4.5 5.5 -0.3 9	 4.5 5.5 -0.3 6
SUPPLY CURRENT (mA): Standby Recessive Driven	 0.300 110	 0.150 18 70	 No Standby 0 20 0 80	 1.5 13 70	 0.200 15 80
MAX SPEED	1 Mb/s	1 Mb/s	500 Kb/s		1 Mb/s
DRIVE CURRENT CAN_H (mA)			–		
MAX VOLTAGE (V) CAN_H, CAN_L	-3 65	-25 25	-5 36		-20 40
VOLTAGES (V) CAN_H CAN_L		 2.75 4.5 0.5 2.25	 – –	 2.75 4.5 0.5 2.25	 3 0.6xVcc 0.4xVcc 2
COMMON MODE RANGE (V)	-7 12	-7 12	-5 12	-7 12	
DIFFERENTIAL VOLTAGE (V) Passive Driven		 -0.5 +0.05 1.5 3.0	 – –	 -0.5 +0.05 1.5 3.0	 -0.5 +0.05 1.5 3.0
PROP DELAY(ns): Transmitter (HS) Receiver (HS)	 100	 80 160	 50 150	 80 145	 50 100
SHORT CIRCUIT CURRENT (mA)		250	150	–	180
INPUT LOAD: Resistance (Ω) Capacitance (pF)		 TBD 68 K TBD	 19 K 70 K 47	 30 K 54 K 20	 – 20 K –
WAVESHAPING (V/µs)	Variable	1)Hi Spd 2)Waveshape 3)Standby	On/off	1)Hi Spd 2)Waveshape 3)Standby	1)Waveshape 2)Standby
BUS TRANSIENTS: ESD Overvoltage	 +/- 2 KV -200 +200 V	 +/- 2 KV	 +/- 2 KV	 +/- 2 KV -150 +100	 +/- 2 KV
OTHER FEATURES		Short Circuit I limit Thermal Shutdown	Short Circuit I limit Thermal Shutdown	Cross Wire Protection Intended for Industrial	3.3 V option 12V & 24V apps

References

[1] Lupini, C. "Advantages of Integrating a Serial Data Link Controller with CPU." Proceedings of the 42nd Vehicular Technology Society Conference, 1992.

[2] Beyeler, B., and Langley, T. "Using the MC68332 TPU to Implement the J1850 Protocol." SAE Paper No. 940137, Society of Automotive Engineers, Warrendale, PA, 1994.

[3] Eisenhart, S., and Baum, D. "Hardware/Software Partitioning for Automotive Multiplex Systems." SAE Paper No. 910711, Society of Automotive Engineers, Warrendale, PA, 1991.

[4] Powers, C. "Cost vs. Performance: Hardware/Software Trade-Off Considerations in Multiplex Device System Design" SAE Paper No. 931806, Society of Automotive Engineers, Warrendale, PA, 1993.

[5] Dais, S., and Chapman, M. "Impact of Bit Representation on Transport Capacity and Clock Accuracy in Serial Data Streams." SAE Paper No. 890532, Society of Automotive Engineers, Warrendale, PA, 1989.

[6] Wollschlager, R., Wellnitz, K., and Frank, R. "Diagnostics and Communications in Vehicle Control Systems Using Smart Power ICs." SAE Paper No. 940627, Society of Automotive Engineers, Warrendale, PA, 1994.

[7] Bartlett, A., and Croft, D. "Tradeoffs in Handling Unwanted J1850 Messages." SAE Paper No. 952579, Society of Automotive Engineers, Warrendale, PA, 1995.

[8] Kiencke, U., and Dais, S. "Application Specific Micontroller for Multiplex Wiring." SAE Paper No. 870515, Society of Automotive Engineers, Warrendale, PA, 1987.

[9] 82527 Serial Communications Controller Architectural Overview. Intel, 1995.

[10] 82527 Serial Communications Controller Data Sheets. Intel, 1995.

[11] MC68HC(7)05X4 Advance Information. Motorola, 1992.

[12] High-Speed CAN ISO Specification 11898.

[13] Peter, C. "Physical Layer for Multiplex Networks." SAE Paper No. 900694, Society of Automotive Engineers, Warrendale, PA, 1990.

[14] Low-Speed CAN ISO Specification 11519-2.

[15] PCA82C250 CAN Controller Interface. Philips, 1994.

[16] SMD Chokes Ensure Electromagnetic Compatibility in Automotive Bus Sytems. Siemens, application note, 1993.

[17] Rall and Sammet. "How to Reduce Noise in Automotive Systems." *EMC Test & Design*, 1994.

[18] EMI Suppression Chokes for Data and Signal Lines. S + M Co., application note, 1995.

Chapter 11

Electromagnetic Compatibility (EMC)

Introduction

This chapter is not intended to be a tutorial in electromagnetic compatibility (EMC) because many texts on the subject ([1], for example) are available. Rather, it is intended only to point out some key EMC issues relating to serial data networks.

Importance of Encoding

Sometimes, "burst" noise (such as that coming through on the radio) cannot be tolerated by the vehicle user. With this in mind, some vehicle manufacturers, such as Lexus, have taken the approach of limiting the amount of bus traffic. Careful attention is paid to bus use and messaging protocol. Unnecessary handshaking and acknowledgments are eliminated. Only the bare-bones messages necessary to convey needed information are sent. This keeps RFI to a minimum.

The type of bit representation (e.g., NRZ, PWM) can have a large effect on RFI [2].

Electromagnetic Compatibility (EMC)

Electromagnetic compatibility (EMC) has the following four main areas of concern:

- Radiated emissions
- Radiated susceptibility
- Conducted emissions
- Conducted susceptibility

A graphical depiction of these is shown in Fig. 11.1.

Radiated Emissions

Radiated emissions are definitely the most discussed area of the four. It is well known that there is less interference after harmonic drop-off, which is roughly calculated as

$$\text{Harmonic drop-off} \Rightarrow 1/(\pi \times \text{edge rate})$$

These are some examples of various edge (rise/fall) rates:

Edge rate:	Harmonic drop-off:
20 ns	16 MHz
10 ns	32 MHz
5 ns	64 MHz
1 ns	320 MHz

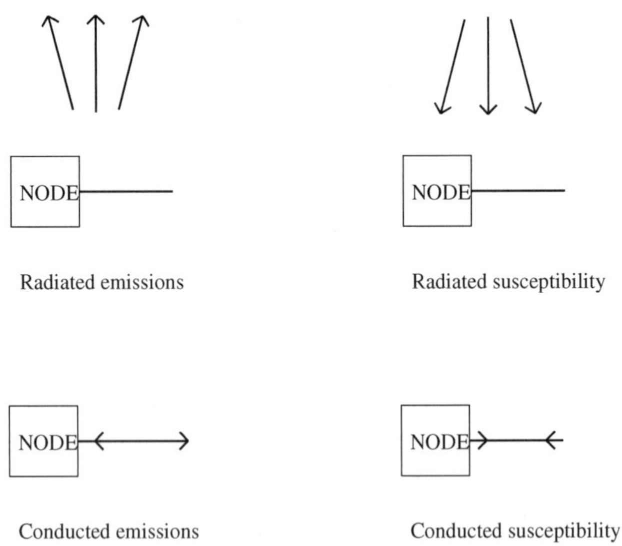

Fig. 11.1 The four EMC areas of concern.

The squarer the signal, the more the spectral content required to make the signal. Radiated emissions depend on harmonic content, the amount of current involved, and the geometry of the return path—not only the clock frequency itself [3]! The switching rate of the bus, however, is usually the number one cause of radiated noise.

As an example of current affecting the emissions, a single-wire bus will emit approximately 3 dB more energy for each 2-volt increase in V_{oh}.

How to Increase Radiated Emissions Performance?

There are many ways to improve the radiated emissions performance of a network [4]. They are not all "cut and dried" because every real-world application is different.

One common improvement is to use differential-mode signaling on two wires. The signal may be current based or voltage based. The main idea is that each wire effectively carries half the voltage or current. The signal on each wire is a mirror image (reflected about their resting or passive state) of the other. The opposite-going nature of these signals tends to cancel each other's harmonics, typically, by ensuring equal and opposite current in the wires, perhaps assisted by a common mode choke. This method is often used with the CAN protocol with excellent results [5–7].

Less than rail-to-rail operation is desirable (small differential) so as to have a better chance at matching the two signals and thus canceling harmonics. This goes for the rise and fall times also. They have to be finely matched (of the order of 1%) for good results. Any common mode current or voltage will show up as poor radiated emissions performance.

Another technique for improving RFI is simply to slow down the bus switching rate. A rough rule of thumb for "speed limit" is as follows :

- About a 20-kHz limit for one wire
- About a 250-kHz limit for a two-wire, unshielded twisted pair
- About a 500-kHz limit for a two-wire, shielded twisted pair

Twisting and/or shielding are commonly used. Twisting cancels the magnetic component of the emissions by having small "loops" instead of one big one (formed by the signal and return). Shielding helps cancel the electric component of the generated field.

As a "last resort," fiber optics is necessary in many cases if the bus must be switched at 500 kHz or more. Then, fiber can theoretically be used out to ∞ MHz.

How to Increase Radiated Susceptibility Performance?

All of the techniques for improving radiated emissions can be used to improve radiated susceptibility. Differential-mode signaling, slow switching speed, twisting/shielding, and fiber optics all help the susceptibility problem.

Proper grounding and termination of the data link is very important so that the bus (or other topology) is not an antenna that is broadcasting, say, WCAN (W Controller Area Network)!

In general, it is expected that some interference will always arise. This is why the error detection and handling schemes are so vital.

Importance of Waveshaping in Single-Wire Systems

An important side note to radiated emissions is the contribution of waveform corners to the total emitted energy. This is especially true for single-wire buses like J1850 VPW. A square wave is certainly not satisfactory, and usually a trapezoidal wave is not, either. What is needed is a higher-order function to dictate a smoothing of the transitions, or corners, of the waveform.

As an illustration of the importance of waveshaping, Fast Fourier Transform (FFT) plots will be shown for a series of waveforms (Fig. 11.2). The FFT for the square wave at 4096 Hz is shown in Fig. 11.3, the FFT for the trapezoidal wave at 4096 Hz is shown in Fig. 11.4, and the FFT for a waveshaped wave at 4096 Hz is shown in Fig. 11.5. Note the different vertical axis scales for each of the FFT plots.

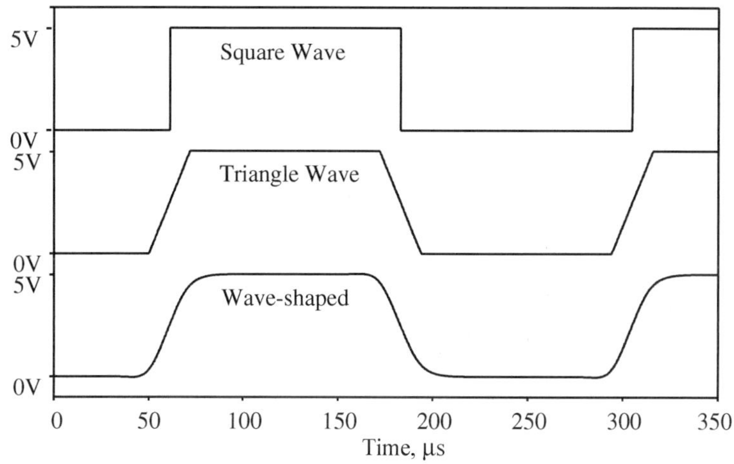

Fig. 11.2 Waveforms for waveshaping comparison.

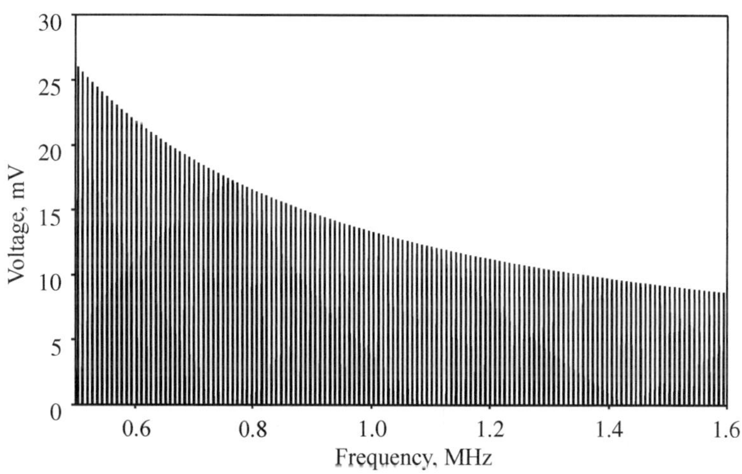

Fig. 11.3 FFT of 4096-Hz square wave.

The relative differences in the narrowband spectral voltage outputs between the square wave and the triangle wave are of the order of 30 to 40 dB. The differences in the narrowband spectral voltage outputs between the triangle

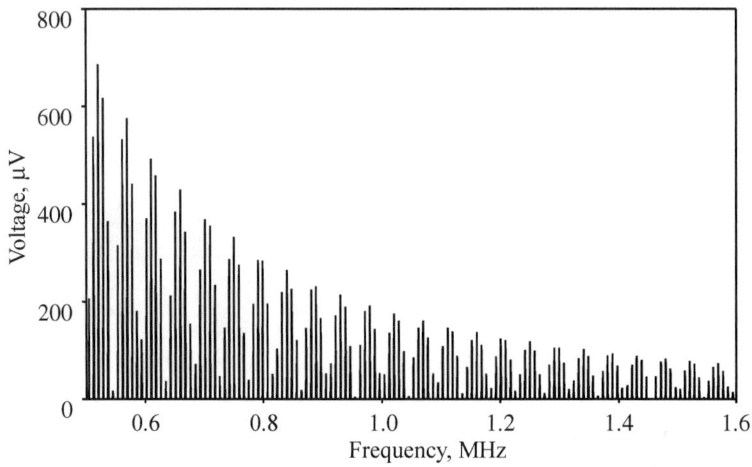

Fig. 11.4 FFT of 4096-Hz trapezoidal wave.

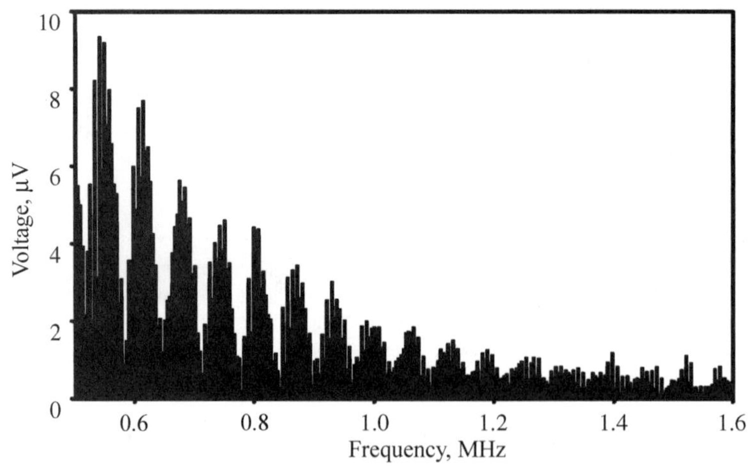

Fig. 11.5 FFT of 4096-Hz shaped wave.

wave and the waveshaped waveform are of the order of 37 to 39 dB. Thus, waveshaping provides significant EMC attenuation in the AM frequency band. The waveshaped output is the only waveform that meets the typical automotive radiated emissions requirements.

Importance of Matching on Dual-Wire Systems

In any two-wire differential communication system, matching of the rise/fall times of the two wires is paramount. In actual implementations, the matching is not perfect, and radiated emissions result [8, 9]. Figure 11.6 shows the ideal case where the rising and falling rates of the two signals match each other during transitions. There is no common mode voltage. In reality, this does not occur. The rise/fall rates of the two wires do not match precisely because of tolerances in the transistors controlling the turn-on/turn-off of the current of voltage drivers, temperature effects, capacitive loading of the bus lengthening fall times more than rise times, and so forth. The resulting common mode voltage or current is the major contributor to RFI.

A mismatch is usually more apparent when the slew rate is larger. It is easier to make the two signals symmetrical if the transitions are faster because there is less time for the rise/decay rates to differ.

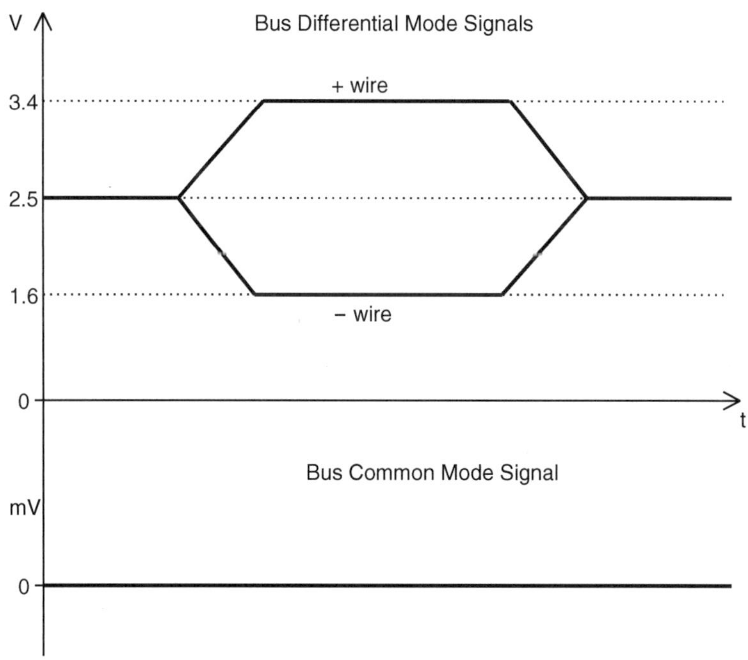

Fig. 11.6 Matched CAN waveform.

Figure 11.7 shows a case where the rise/fall times do not match. The falling portion of each signal is longer than the rising portions. There is a corresponding common mode voltage. This results in large radiated emissions.

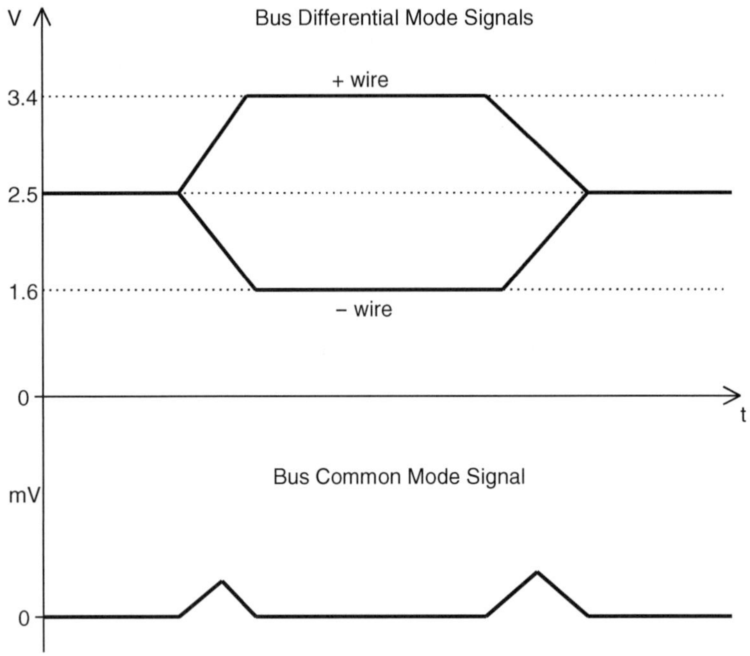

Fig. 11.7 Unmatched CAN waveform.

References

[1] Ott, H. "Noise Reduction Techniques in Electronic Systems." 2nd Ed., Wiley-Interscience, 1988.

[2] "The Impact of Bit Representation on Noise in Automotive Networks." Unpublished Bosch document.

[3] "The Hidden Schematic." Intel *Common Ground* newsletter, 1993.

[4] Rall and Sammet. "How to Reduce Noise in Automotive Systems." *EMC Test & Design*, 1994.

[5] McLaughlin, R. "EMC Susceptibility Testing of a CAN Car." SAE Paper No. 932866, Society of Automotive Engineers, Warrendale, PA, 1993.

[6] "CAN Bench Test for Radiated Susceptibility." Unpublished Bosch document.

[7] "Electromagnetic Radiation and Susceptibility of the CAN Bus." Unpublished Bosch document.

[8] Eisele, H., and Lutjens, H. "EMC Measures for Class C Communication Systems Using Unshielded Cable." SAE Paper No. 950298, Society of Automotive Engineers, Warrendale, PA, 1995.

[9] Lupini, C., and Eisele, H. "Employing Unshielded Cable for Class C Multiplexing." Proceedings of 1st International CAN Conference, 1994, www.can-cia.de.

Chapter 12

What Is a Standard?

Introduction

This chapter gives some insight into the major world standards bodies and their published documents.

ISO

The International Organization for Standardization (ISO) is a voluntary, nontreaty (i.e., no written contract among members guaranteeing adherence to the standards) group with associated arms from various countries. These are some affiliated members of ISO:

- United States: American National Standards Institute (ANSI)
- United Kingdom: British Standards Institute
- Canada: Standards Council of Canada

ISO has various working groups that work under an umbrella of committees. The applicable serial data group is ISO/TC22/SC3/WG1/TFx. This is Technical Committee (TC) number 22, which deals with road vehicles. Subcommittee (SC) number 3 handles electrical and electronic equipment, and Work Group (WG) number 1 is for serial data communication.

Within the Work Groups are Task Forces (TFx). Several of the WG1 Task Forces are discussed elsewhere in this book. They include TF1 for Diagnostic Services, TF2 for Diagnostics on CAN, TF3 for 9141-2 verification, TF4 for Truck & Bus Interface, and TF5 for SAE J-spec conversion.

Some ad hoc task forces are usually active. At this time, there are ad hoc task forces for CAN conformance, tachograph systems, and extended data link security.

To obtain specifications, contact the following:

Central Secretariat
International Organization for Standardization
1, Rue de Varembe
CH-1211 Geneva Switzerland
41-22-34-12-40
www.iso.org

American National Standards Institute
1430 Broadway
New York, NY 10018
(212) 354-3471
www.ansi.org

Copies of many of the ISO documents are available now through SAE.

Some applicable ISO standards that will be discussed in succeeding chapters are introduced in the next several subsections.

ISO 9141-2

- Specifies requirements for a vehicle data communications network
- Alternate protocol allowed by OBD-II

ISO 11519

- State of multiplexing in Europe (CAN, VAN, J1850 are now ISO standards)

IEEE

The Institute of Electrical and Electronic Engineers (IEEE) is an industry group known for many prominent standards. One of the most notable is IEEE 802 LAN, of which "Ethernet" is a member protocol. Another is IEEE 488, a popular instrumentation bus standard.

To obtain specifications, contact

IEEE Computer Society
1109 Spring Street, Suite 300
Silver Spring, MD 20910
(301) 589-3386
www.ieee.org

SAE International

SAE International (formerly the Society of Automotive Engineers, or SAE) is the main group in which we are interested. Although it publishes a wide variety of documents relating to all areas of ground vehicles and aerospace, it has a large library of multiplex-related information.

SAE publishes three main types of documents relating to standards. The first of these is an Information Report (IR). Basically, anyone can write an information report about any subject. It is simply an "FYI" on some aspect of automotive or aerospace engineering that the author would like to disseminate. A Recommended Practice (RP) is the beginning of a standard. It is the working document from a task force or committee charged with developing a new standard. Once an RP has been balloted and approved by the committee members, the RP becomes a Standard.

To obtain specifications, contact

SAE International
400 Commonwealth Drive
Warrendale, PA 15096-0001
www.sae.org
(724) 776-4970

The following is a brief overview of some relevant SAE documents that are discussed in other chapters of this book.

SAE J1850

J1850 specifies requirements for a vehicle data communications network. It is one of the acceptable OBD-II data buses. J1850 comes in two main flavors: a two-wire 41.6-kb/s PWM and a one-wire 10.4-kb/s VPW. A compliance document exists (J1699) with verification procedures.

SAE J1979

Specifies CARB/EPA functions and associated messages.

SAE J2190

Gives diagnostic and malfunction messages for "fixing" vehicles. It goes beyond the emissions-related messages of J1979. Although voluntary, most vehicle manufacturers are implementing J2190 modes.

SAE J1962

J1962 is the requirement for the new 16-pin common underdash diagnostic connector. Almost all new vehicles sold in the United States beginning in the 1996 model year have this connector. The official name of this connector is the data link connector (DLC).

SAE J1978

This document gives the specifications for the OBD-II scan tool. A companion spec is the J2300 compliance test for OBD-II scan tools.

SAE J2178

J2178 specifics the basic message format that is used for OBD-II communications. It has the following four parts:

- Part 1—Message header formats and physical node addresses
- Part 2—Message parametric data

- Part 3—Message assignments for the single-byte header format
- Part 4—Message assignments for the three-byte header format

Others

As indicated earlier, the SAE provides a vast array of automotive technical documentation. Several other electrically related specifications are briefly listed next, in ascending numerical order.

SAE J1113

Electromagnetic Susceptibility Measurement Procedures for Vehicle Components

SAE J1211

Electronic Equipment Design Recommended Procedure

SAE J1213

Glossary of Terms For Vehicle Networks

SAE J1547

Electromagnetic Susceptibility Measurement Procedures for Common Mode Injection

SAE J1587

- Joint SAE/TMC (Truck Maintenance Council) Electronic Data Interchange Between Microcomputer Systems in Heavy Duty Vehicle Applications (NOV90)

- Specifies message formats for use with J1708 physical layer

SAE J1699

Compliance Tests and Test Methods for SAE J1850

SAE J1708

- Serial Data Communications Between Microcomputer Systems in Heavy Duty Vehicle Applications
- Specifies mainly the physical layer (RS-485 based) for J1587 and J1922

SAE J1879

General Qualification and Production Acceptance Criteria for Integrated Circuits in Automotive Applications

SAE J1922

- Powertrain Control Interface for Electronic Controls Used In Medium and Heavy Duty Diesel On-Highway Vehicle Applications
- Message formats for data passing between engine, transmission, ABS, and so forth

SAE J1930

Electrical/Electronic Systems Diagnostic Terms, Definitions, Abbreviation, and Acronyms

SAE J2008

Recommended Organization of Vehicle Service Information

SAE J2012

Diagnostic Codes

SAE J2205

Expanded Diagnostic Protocol for OBD-II Scan Tools

SAE J2300

Conformance Test Procedures for OBD-II Scan Tools

Figure 12.1 is an example of a flowchart that shows how J2201 specifies the way an OBD-II scan tool learns which data bus is in the subject vehicle.

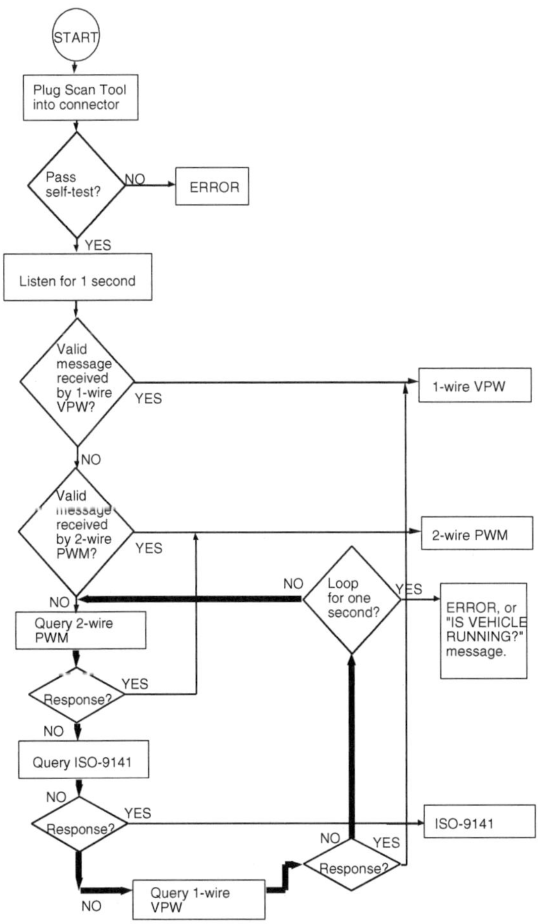

Fig. 12.1 J2201 flowchart.

Figure 12.2 shows an example of the hardware implementation of a J2201 scan tool interface. There are several options, but this one is common.

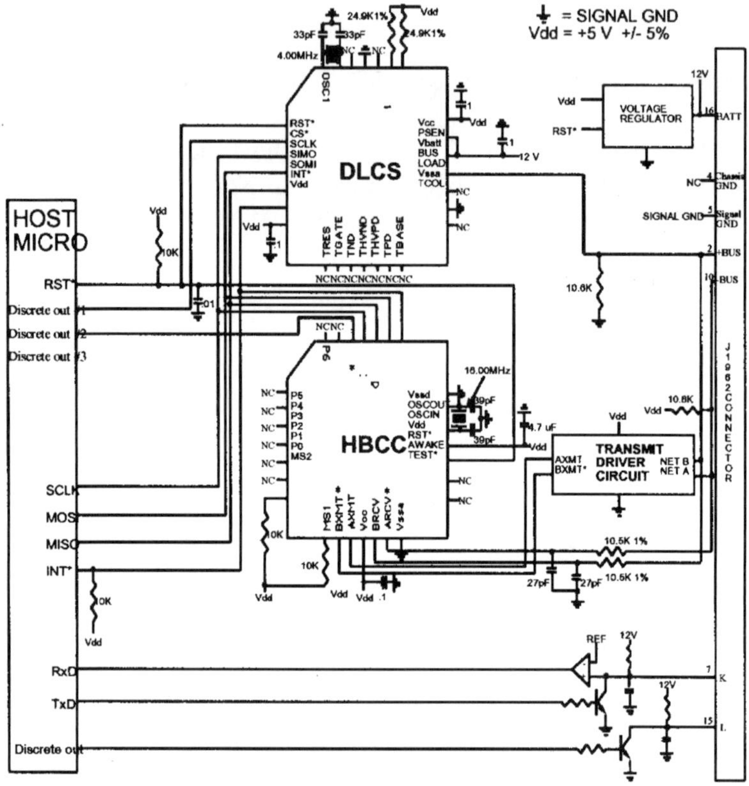

Fig. 12.2 Block diagram of a typical J2201 scan tool interface.

J1962/ISO 15031-3

In the United States, J1962 is a very important automotive electrical standard. The corresponding ISO version is ISO 15031-3. This specification finally harmonizes the in-car, underdash diagnostic connectors of most vehicle makers. A layout of this connector is shown in Fig. 12.3, and the corresponding pin assignments by manufacturer are listed in Table 12.1.

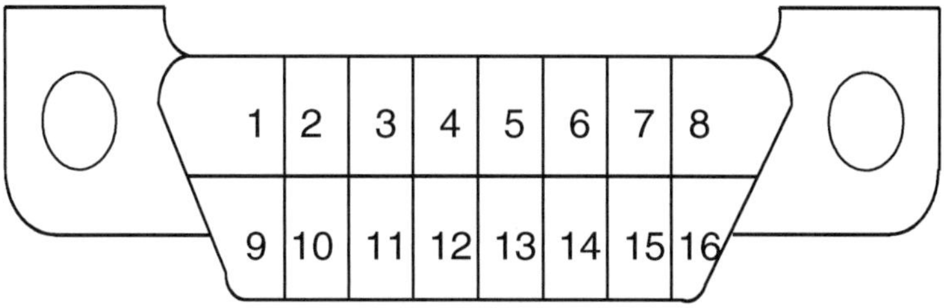

Fig. 12.3 Representation of a J1962 diagnostic connector.

TABLE 12.1
J1962 PIN ASSIGNMENTS

Terminal	SAE	General Motors	DaimlerChrysler	Ford
1	Discretionary	UART (secondary)	Keyless entry enabler	Ignition control
2	J1850 +	Class 2	J1850 +	SCP +
3	Discretionary	Ride control enabler	CCD +	
4	Chassis ground	Chassis ground	Chassis ground	Chassis ground
5	Signal ground	Signal ground	Signal ground	Signal ground
6	*(CAN H line)**	*(CAN H line)**	*(CAN H line)**	*(CAN H line)**
7	ISO 9141-2 K	ISO 9141-2 K	ISO 9141-2 K	ISO 9141-2 K
8	Discretionary	Keyless entry enabler	Switched Vbatt	Trigger signal in
9	Discretionary	UART (primary)		Switched Vbatt
10	J1850 –	Unused	J1850 –	SCP –
11	Discretionary	Steering control	CCD –	
12	Discretionary	ABS, CCM enabler		Flash EEPROM
13	Discretionary	SIR diagnostics		Flash EEPROM
14	*(CAN L line)**	*(CAN L line)**	*(CAN L line)**	*(CAN L line)**
15	ISO 9141-2 L	ISO 9141-2 L	ISO 9141-2 L	ISO 9141-2 L
16	Unswitched Vbatt	Unswitched Vbatt	Unswitched Vbatt	Unswitched Vbatt

* *De facto* use by manufacturers.

Chapter 13

Class A Protocols

Introduction

This chapter discusses several standardized, or at least well-known, Class A serial data link protocols and systems. These protocols are usually of the same physical type and are collectively referred to as the "UARTs" (universal asynchronous receiver/transmitters).

A series of SAE Information Reports do a good job of introducing the basis for Class A multiplexing [1–4]. Reference [5] is also an interesting SAE report about a DaimlerChrysler sensor and control (CSC) bus. To read about several Class A applications, refer to [6–8]. Note that [8] eventually became the Toyota "BEAN" protocol [9].

Class A implementations are supposed to reduce vehicle wiring to a greater extent than the other classes. Most of these applications are "body" type, which includes such nodes as mirrors, door locks, seat controls, and so forth. Although wiring can be reduced as much as 20 to 50%, the cost usually increases [10, 11].

It is also important to remember that we are trying to reduce wire count and also terminal count, splice count, wire length, weight, and so forth.

SAE J1708

The first of the UART-based protocols that should be mentioned is part of a group used in heavy-duty trucks [12]. J1708 plus J1587 or J1708 plus J1922 have been used since the 1980s and will continue for several more model years. The CAN protocol, specifically J1939, is due to replace these earlier UART-based systems. Multiplexing in monitoring systems has been used for many years in the heavy-duty freight industry [13–17].

J1708 is titled "Serial Data Communications Between Microcomputer Systems in Heavy Duty Vehicle Applications—JAN1986." It was developed by the SAE Truck and Bus committee and is widely used by the major over-the-road freight trucking manufacturers. This specification is very limited in scope and covers only the physical layer and a small part of the data link layer.

Electrical characteristics are described, and source addresses and message identifiers (MIDs) are defined. The basic unit of information transfer is a "character" in the usual UART format (see Fig. 13.1).

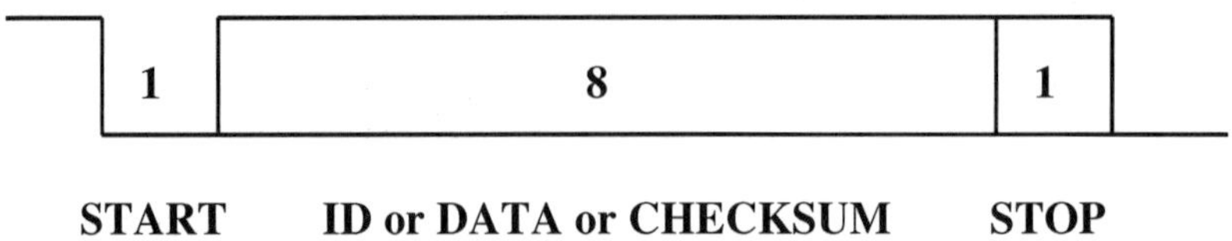

Fig. 13.1 J1708 character format.

A character can be a message ID (MID), undefined information, or a checksum.

Multiple successive characters make up a message. The basic J1708 message format is shown in Fig. 13.2.

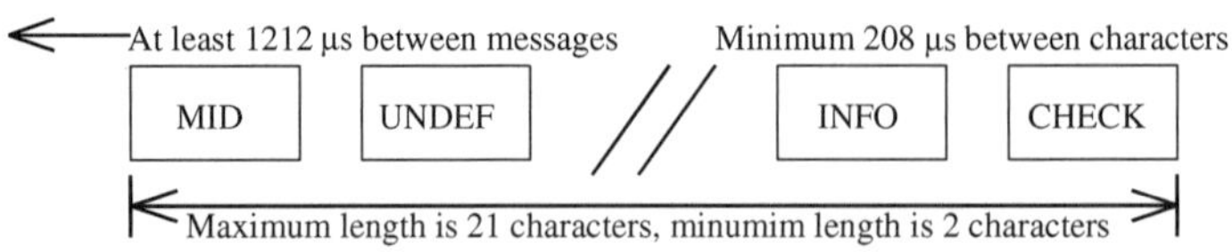

Fig. 13.2 J1708 message format.

Messages are limited to 20 characters by J1708. The bus medium is a two-wire differential that is actually a modified RS-485 link. Encoding and data rate are the usual NRZ at 9600 b/s. As many as 20 nodes can be connected, with a maximum bus length of 40 meters. The types of applications usually used for J1708 are proprietary service tools and proprietary products. This was a major problem with these older Truck and Bus standards, in that they gave a lot of leeway for tweaking. Each vehicle manufacturer made up its own message structure, which was incompatible with that of every other manufacturer.

J1708 forms the basis for two types of systems. It is combined with either J1587 or J1922.

SAE J1587

The official title of J1587 is "Electronic Data Interchange Between Microcomputer Systems in Heavy Duty Vehicle Applications—JAN1988" [18]. This document fills in some of the holes in J1708 and adds some data link layer specifications. It is always used with J1708.

J1587's character format is that specified in J1708 (Fig. 13.1). The message format builds on that given in J1708 and defines blocks (Fig. 13.3).

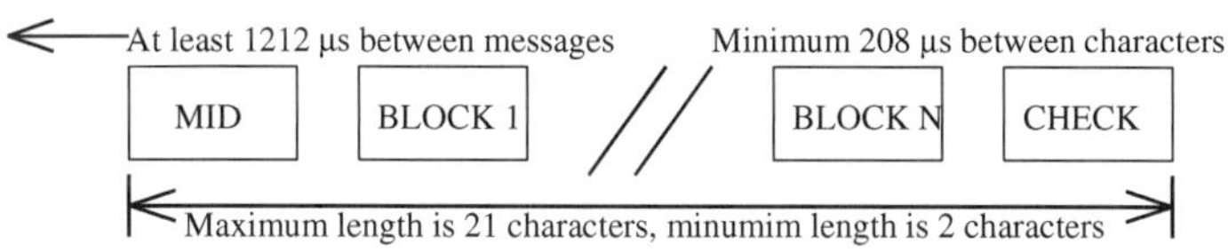

Fig. 13.3 J1587 message format.

There are two block types, commands and labels. Commands can be either global (request for data) or subsystem specific (also a request for data). Labels are functional identifiers indicating revolutions per minute, miles per hour, and so forth. Block formats are shown in Table 13.1.

Within a J1587 application, characters can be parameter IDs (PIDs), message IDs (MIDs), data, message length, or a checksum. Unlike J1708, J1587 predefines some messages. For example, PID = 90 indicates revolutions per minute. J1587 plus J1708 was, and still is, used for many general purpose information-sharing and diagnostic applications.

TABLE 13.1
J1587 BLOCK FORMATS

PID	Block Format
0–127	PID . . . DATA
128–191	PID . . . DATA . . . DATA
192–253	PID . . . LENGTH . . . DATA . . . DATA
254	PID . . . MID . . . PROPRIETARY INFO
255	Reserved for future expansion

The other companion for J1708 is J1922.

SAE J1922

J1922's title is "Powertrain Control Interface for Electronic Controls Used in Medium Duty and Heavy Duty Diesel On-Highway Vehicle Applications—NOV1989" [19]. It was also developed by the SAE Truck and Bus committee and is used for applications different from J1708 plus J1587. J1922 is more specific for powertrain usage. It is closely related to J1587 but adds "control modes," which are speed, torque, speed and torque limit, and torque limit. Like J1587, it relies on J1708 for physical layer definition.

J1922's character format is that specified in J1708 (Fig. 13.1). The message format is a little different from J1708 and defines data fields (Fig. 13.4).

Character types are as in J1708, but J1922 adds definitions for data characters. One predefined message is 22 characters long (technically, it violates

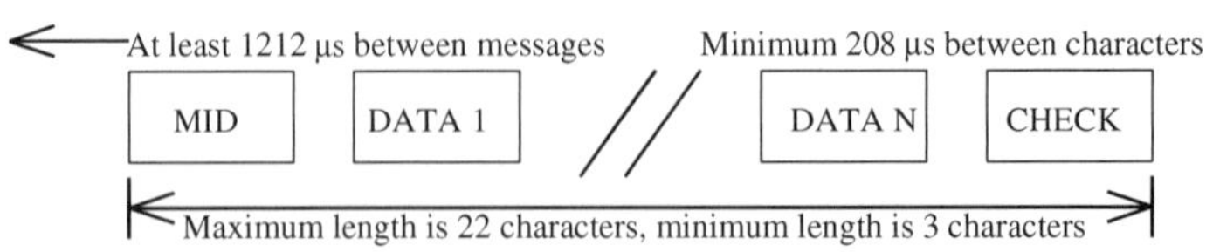

Fig. 13.4 J1922 message format.

J1708). All messages are predefined and standardized and thus are interchangeable among manufacturers.

J1922 limits the node count to four specific units: engine, transmission, ABS/traction, and retarder. The maximum bus throughput is specified as 70%.

CCD (C^2D)

DaimlerChrysler collision detection (CCD or C^2D) is a proprietary bus protocol developed by DaimlerChrysler in the mid-1980s. Public documentation on CCD can be found in two SAE technical papers [20, 21] and an SAE Information Report [22].

The CCD message format, shown in Fig. 13.5, is composed of characters separated by some idle time.

ID	< 10 bits	Data 1	< 10 bits	Data 2	< 10 bits	Data N	< 10 bits	CRC	≥ 10 bits
	IDLE		**IDLE**		**IDLE**		**IDLE**		**IDLE**

Fig. 13.5 CCD message format.

Characters are the usual UART 10-bit format—start bit, eight info bits, and one stop bit. The first character, the functional message ID, is unique (256 possible). Middle characters are data bytes (up to five), and the last character is a CRC (optional).

Other major CCD characteristics are that it is a two-wire twisted pair bus, NRZ encoded, and operates at a data rate of 7812.5 b/s. This speed was chosen for compatibility with clock frequencies that DaimlerChrysler used then.

This was a heady time for automotive multiplexing, and the major players were giddy with excitement over what they saw as a budding new technology that would transform automotive electronics. DaimlerChrysler saw an opportunity to lead in this area and wanted to develop a new serial data protocol. Hence, CCD was born.

CCD would figure into the government's birth of the OBD-II regulations (see Chapter 4).

DaimlerChrysler was part of the development of the present classification of serial data link protocols into Classes A, B, and C. Its take on the classes about 1986 was described as follows: Class A encompassed smart (power) sensors and actuators for which the market was not yet ready. Class C applications would require fiber optics or some other expensive medium and was a way down the road. Class B, or general intermodule communications, was where DaimlerChrysler believed the current work should be directed. Out of this work was born CCD (which could be more correctly categorized as Class A because of its speed).

These were the original requirements for CCD:

- Open architecture
- Capable of Class A activities also
- Message length less than 6 bytes
- Message latency between 10 and 100 ms
- Minimal RAM and ROM support
- Efficient bit encoding technique and message format
- Easy interface to host microprocessor
- Hardware capable of integration onboard a microprocessor
- Good data integrity and reliability
- Reliable physical layer to shorts, and so forth
- Network should be usable for diagnostics and last for more than five years

The final CCD protocol supported collision detection but not resolution (transmitters would have to retry).

Physical layer aspects of CCD were a simple differential current drive over an unshielded twisted pair in a bus topology (see right side of Fig. 13.6).

The microprocessor interface was a standard UART (SCI). Custom hardware was used for the arbitration, filtering, timing logic, and so forth (Fig. 13.6). The first manifestation of this hardware was a chip by Harris Semiconductor called the CDP68HC68S1. This chip was not commercially available. It had

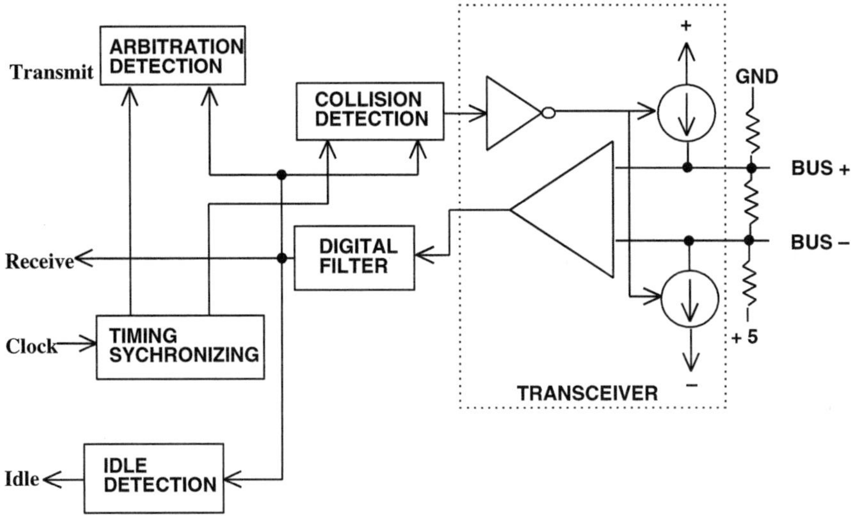

Fig. 13.6 Block diagram of CCD bus interface.

fewer than 600 gates, and the CRC calculation was left up to the host microprocessor. Both SPI and SCI interfaces for the host were provided.

DaimlerChrysler has replaced all CCD implementations with J1850 one-wire [23] and is now replacing those with CAN and other protocols.

ISO 9141

ISO 9141 is a European-based automotive diagnostic specification that has been around for a long time (at least 20 years). It is practically a standard UART interface with some extra specifications. Many of the details are left up to the user, so most vehicle manufacturers put their own "flavor" of higher layers on top of 9141. ISO 9141 was intended for point-to-point diagnosis between the engine or powertrain controller in the vehicle and an offboard tester. The basic communication protocol is specified so that a test node establishes the fine points of the dialog with the vehicle.

The protocol uses a basic UART type of character (Fig. 13.7).

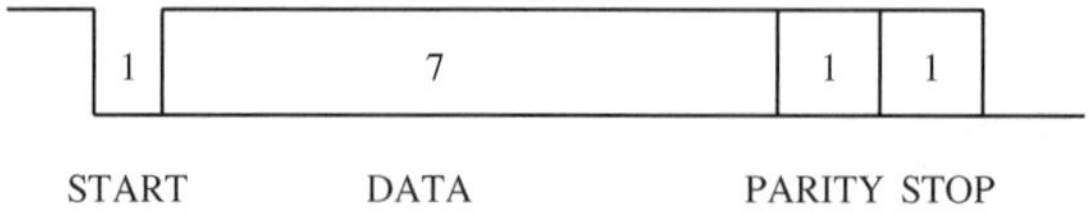

Fig. 13.7 ISO 9141 character format.

Characters convey address or data information. How the addresses are defined is specified in higher-layer documents, such as the "keyword" family (discussed later). One or two (optional) bus lines are supported. The required signal, the K line, carries bidirectional address and data information. An optional L line is basically a redundant unidirectional line from the tester that carries the same address information as the K line. The user defines the data rate, but it is limited to about 10 kb/s.

An ISO 9141 electrical interface can easily be made from discrete components. There are also ICs available, such as the Siliconix Si9241 (K line only) and Si9243 (K and L lines), which are economical and in an eight-pin package.

ISO 9141-2

ISO 9141-2 is the name given to the version of ISO 9141 used to satisfy OBD-II requirements in the United States. It is sometimes referred to as the "European version of J1850." CARB accepted ISO 9141-2 for OBD-II at the request of European vehicle manufacturers.

This action was significant because it "harmonized" a U.S. and European (and Asian) standard serial data link. Europe and Asian vehicle manufacturers use ISO 9141-2 almost exclusively to satisfy OBD-II. At least one U.S. vehicle maker also uses it—DaimlerChrysler in the Dodge Neon, for example.

ISO 9141-2 requires the message structure specified in J1979, and the K and L lines are reserved in the J1962 diagnostic connector. Address bytes are transferred at 5 b/s, and data bytes travel more quickly at 10.4 kb/s. The L line is optional in the protocol.

Like the basic uses of ISO 9141, 9141-2 is for use only between the engine controller and an offboard scan tool.

At least one proposal has been made to upgrade ISO 9141 and ISO 9141-2 to a network protocol with more than two nodes [24]. At this time, no one has attempted to do this for a mass production application.

Keyword 2000

A little explanation of a keyword is in order at this point. Many keyword protocols exist and are based on ISO 9141. They are essentially flavors of ISO 9141 used by various vehicle manufacturers. A keyword is 2 bytes that are transferred by the tester/master at 5 b/s over the K (and L if present) line. The onboard node responds with 55h at the required faster speed so the master can verify the bit rate. The onboard node also repeats the keyword back to the master node. It is the keyword that indicates message header length to be used, bit rate, and so forth. Many keywords have been defined that are manufacturer-specific and proprietary. Keyword 2000 is an exception to this.

Keyword 2000 is considered a follow-up development effort to ISO 9141-2. The responsible ISO work group is ISO 14230 which is part of ISO/TC22/SC3/WG1/TF1. That group is working to make Keyword 2000 the standardized protocol for ISO 14229 (Diagnostic Services).

Surprisingly, many vehicle manufacturers are already using Keyword 2000. One of the reasons is that it greatly expands the usefulness of ISO 9141 by providing for both physical and functional addressing of messages.

Actually, Keyword 2000 specification comprises a family of keywords (i.e., 2000, 2001, 2002). They all communicate at 10.4 kb/s. An additional tidbit of information concerning Keyword 2000 is that the number 2000 really is not the character (in decimal or hexadecimal) sent by the tester. The actual word used is 8FD0h.

GM UART

The GM "UART" bus (also called "ALDL") is an 8192-b/s, half-duplex, NRZ universal asynchronous receiver/transmitter type of link. A typical system consists of one master device, one or more slave devices, and data link connector

(DLC), which was formerly called the assembly line diagnostic link (ALDL). During normal communications, the master schedules all traffic on a continuous, repetitive basis. If the system is set up as a broadcast remote, all slave devices listen to all messages but do not respond to any of them. If, however, the system is set up in a polled fashion, the master sends a specific message to each slave in turn, after which an immediate response is expected. In the test/service mode of operation, an external device connected to the DLC usually serves as the master device on the bus.

The UART bus has been used since the 1980 model year but is gradually fading out. It was originally forecast that it would disappear by the 2000 model year but was still used in the 2002 model year. The UART has several major strong points: ultralow cost and a low maintenance and failure rate, it is well understood, and the service community is very familiar with it.

Nodes on a typical UART link include the powertrain controller and possibly an ABS and/or body computer. An offboard tester or programmer constitutes the master node. Some high-end systems, such as Cadillac, contain as many as seven nodes with the powertrain controller as the master.

GM Entertainment and Comfort (E&C)

The GM "Entertainment and Comfort" (E&C) bus has been used for many years. Developed for audio and HVAC applications, it is a CSMA/CD PWM type of link. It operates at 1 kb/s and generally carries control and status information. The bus idles high (pulled up to battery voltage) and is actively pulled low (to ground) by a transmitting device. When the line has remained inactive for a specified time, any node may begin transmitting onto the bus. While transmitting, each node continues to sample the line to ensure that what it is transmitting is actually appearing on the bus. A zero bit is transmitted by holding the bus low for a "short" time; holding the bus low for a "long" time transmits a "one" bit. Therefore, the E&C bus is "ones dominant" for arbitration among multiple transmitters.

As with the UART bus, E&C is still used today and will continue for at least several more model years, but it is fading fast.

Local Interconnect Network

The local interconnect network (LIN) protocol is gaining strength and is likely to become the world standard for Class A multiplexing [25]. The increasing features and complexity of today's automotive architectures are becoming increasingly difficult to manage. Each new innovation typically requires additional mechanical actuators and associated electrical controllers. The sheer numbers of black boxes and wiring are not being limited by features or cost but by the inability to assemble them into a vehicle. A new architecture is required that will support the ability to add new features and also enable the vehicle assembly plants to assemble and test each subsystem easily. One such architecture is a distributed multiplex arrangement that reduces the number of wires while enabling flexibility and expandability. Previous versions have had to deal with issues such as noise immunity at high switching currents. The LIN bus with its low cost and rail-to-rail capability may be the key enabling technology to make multiplexed architecture a reality. The LIN protocol is covered in depth in the LIN specifications. That and more information is available at www.lin-subbus.org.

References

[1] SAE J2057/1 Class A Application/Definition. Society of Automotive Engineers, Warrendale, PA.

[2] SAE J2057/2 Class A Multiplexing Actuators. Society of Automotive Engineers, Warrendale, PA.

[3] SAE J2057/3 Class A Multiplexing Sensors. Society of Automotive Engineers, Warrendale, PA.

[4] SAE J2057/4 Class A Multiplexing Architecture Strategies. Society of Automotive Engineers, Warrendale, PA.

[5] SAE J2058 DaimlerChrysler Sensor and Control Bus. Society of Automotive Engineers, Warrendale, PA.

[6] Sakagami, A., Fujiki, N., Mitimura, K., Kimura, T., Matsumoto, T., Nomura, K., and Oho, S. "Development of an On-Board Class A Local Area Network System." SAE Paper No. 920229, Society of Automotive Engineers, Warrendale, PA, 1992.

[7] Furuichi, K., Ishida, K., and Akashi, K. "An Implementation of Class A Multiplex Application." SAE Paper No. 920230, Society of Automotive Engineers, Warrendale, PA, 1992.

[8] Isobe, T., Honda, H., Uehara, S., and Akiyama, S. "A Low-Speed In-Vehicle Network for Body Electronics." SAE Paper No. 920231, Society of Automotive Engineers, Warrendale, PA, 1992.

[9] Honda, H., Uehara, S., Sakai, K., Akatsuka, T., and Akiyama, S. "Body Electronics Area Network (BEAN)." SAE Paper No. 970297, Society of Automotive Engineers, Warrendale, PA, 1997.

[10] Ishida, K., Akashi, K., and Kato, T. "Verification of the Effect of Class A Multiplex Application on the Wiring System." SAE Paper No. 910471, Society of Automotive Engineers, Warrendale, PA, 1991.

[11] Maruyama, M., Tani, M., and Terada, T. "A New Approach to Vehicle Interior Control." SAE Paper No. 910472, Society of Automotive Engineers, Warrendale, PA, 1991.

[12] SAE J1708. Society of Automotive Engineers, Warrendale, PA.

[13] Riley, D., and Onesti, R. "The Importance and Evolution of the Standardized Serial Data Link to the Truck OEM." SAE Paper No. 852325, Society of Automotive Engineers, Warrendale, PA, 1985.

[14] Bishel, R. "Maximizing Utilization of the SAE J1708 Serial Data Communication Link." SAE Paper No. 861069, Society of Automotive Engineers, Warrendale, PA, 1986.

[15] Stepper, M. "Data Link Overview for Heavy Duty Vehicle Applications." SAE Paper No. 902215, Society of Automotive Engineers, Warrendale, PA, 1990.

[16] Munch, J. "Evolution of Vehicle Based Monitoring Systems." SAE Paper No. 933069, Society of Automotive Engineers, Warrendale, PA, 1993.

[17] Luebbering, B., and Smith, V. "Engine Electronics Technology." SAE Paper No. 932404, Society of Automotive Engineers, Warrendale, PA, 1993.

[18] SAE J1587. Society of Automotive Engineers, Warrendale, PA.

[19] SAE J1922. Society of Automotive Engineers, Warrendale, PA.

[20] Miesterfeld, F. "DaimlerChrysler Collision Detection (C^2D)—A Revolutionary Vehicle Network." SAE Paper No. 860389, Society of Automotive Engineers, Warrendale, PA, 1986.

[21] Madden, W. "DaimlerChrysler Collision Detection (C^2D^{TM}) Bus Interface, Integrated Circuit User Manual." SAE Paper No. 880586, Society of Automotive Engineers, Warrendale, PA, 1988.

[22] SAE J1587, Collision Detection Serial Data Communications Multiplex Bus. Society of Automotive Engineers, Warrendale, PA.

[23] Miesterfeld, F., and Luitje, W. "Full Vehicle Multiplexing at DaimlerChrysler." SAE Paper No. 950033, Society of Automotive Engineers, Warrendale, PA, 1995.

[24] Samual, J., and Pischke, J. "An Arbitration Mechanism on ISO 9141-2." SAE Paper No. 950041, Society of Automotive Engineers, Warrendale, PA, 1995.

[25] Perisho, R., Ewbank, S., Kleja, F., and Lupini, C. "LIN Bus and Its Potential for Use in Distributed Multiplex Systems." SAE Paper No. 2001-01-0072, Society of Automotive Engineers, Warrendale, PA, 2001.

Chapter 14

Class B Protocols

Introduction

Class B protocols are generally those in the speed range of 10 to 125 kb/s or so. It also encompasses those protocols that are used primarily for general purposes such as diagnostics and going-down-the-road status operations.

Vehicle Automotive Network

The vehicle area network (VAN) protocol was developed by Renault and Peugeot with assistance from Siemens [1, 2]. The project also had the financial backing of the French government. VAN was standardized in ISO 11519-3 [3].

Renault, Peugeot, Volvo, and Fiat principally used VAN. Applications included powertrain and body/chassis functions. VAN has now fallen out of favor and is being replaced by CAN.

The protocol incorporated a unique encoding scheme called enhanced Manchester. Each message nibble is encoded with three NRZ bits followed by an L-Manchester bit to guarantee an edge every four bits for resynchronization. It can be likened to an "NRZ-4" technique with bit stuffing, but stuffed bits are not necessary. Enhanced Manchester does complicate the transmitter and receiver encoding/decoding circuitry. A diagram illustrating enhanced Manchester is shown in Fig. 14.1.

VAN was applied over a two-wire twisted pair bus structure with as many as 20 nodes. No maximum network length was specified. Although no data rate was standardized, 125 kb/s or 250 kb/s was typically used. Overhead (nondata bits/data bits) is about 41%.

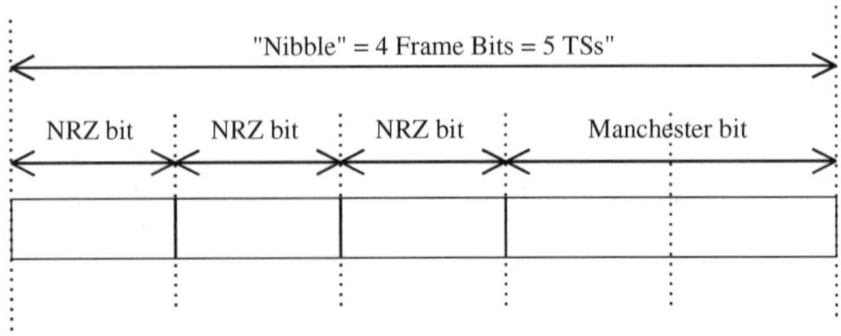

Fig. 14.1 Enhanced Manchester encoding.

The basic message format is shown in Fig. 14.2.

SOF	IDEN	COM	DAT	FCS	EOD	ACK	EOF
1 symbol (10 TS)	12 bits (15 TS)	4 bits (5 TS)	0 - 28*8 bits (0 - 630 TS)	15 bits (18 TS)	(2 TS)	(2 TS)	(8 TS)

Fig. 14.2 Basic VAN message structure.

The COM (command) field decoding is shown in Fig. 14.3.

EXT	RAK	R/W	RTR

Fig. 14.3 VAN message command field.

- EXT bit:

  The extension (EXT) bit is reserved at this time, but must be sent recessively ("1").

- RAK (request acknowledge) bit:

  The one-sending module is requesting an in-frame response (IFR). The responding node sets the ACK (acknowledge) field to a one.

The zero-sending module is not requesting an in-frame response (IFR).

- R/W (read/write) bit:

  The one message sent is a read request.

  The zero message sent is a write request.

- RTR (remote transmit request) bit:

  The one frame contains no data and is a request for a remote transmission.

  The zero frame contains data and is not a request for a remote transmission.

VAN supports several message types, including data, remote (request), and response.

VAN provided some important features for which no license was required to produce ICs: powerful in-frame response, 28-byte messages, and reportedly low RFI.

Several ICs were developed to support VAN applications. These will be discontinued as all users of VAN switch to CAN.

Automobile Bus

ABUS (asynchrone bitserielle unversalschnittstelle), sometimes called automobile bus, was developed by Volkswagen and competed with CAN and VAN for European acceptance. It never became a standard due to ISO's refusal to document ABUS. Hence, all new development has ceased, and new applications will adopt CAN.

The relatively simple message format (although it still had about 48% overhead) is shown in Fig. 14.4.

Message types were data and command. A two-wire twisted pair or a single wire in a bus structure at 500 kb/s was specified. Encoding was NRZ. The maximum number of nodes was not specified.

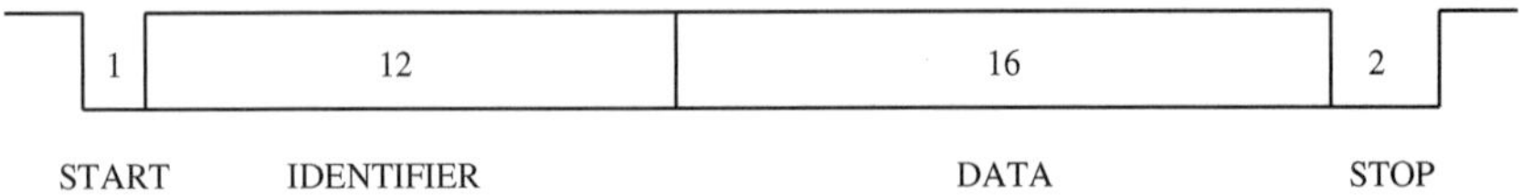

Fig. 14.4 ABUS message format.

J1850

SAE J1850 [4] is the dominant Class B serial data standard in the United States. Ratified by GM, Ford, and DaimlerChrysler on February 1, 1994, it had been an SAE Recommended Practice for seven years.

SAE J1850 describes primarily requirements at the data link and physical layers. There are two main alternatives in implementing J1850, as seen in Fig. 14.5. These differences can be described as 41.6 kb/s, PWM, two-wire, and 10.4 kb/s, VPW, one-wire. Ford is the user of the two-wire form, and GM and DaimlerChrysler use the one-wire. There is a higher layer difference via J2178 that further separates GM and DaimlerChrysler. GM uses only 3-byte message headers, and DaimlerChrysler uses 1-byte and 3-byte headers in approximately a 35–65 split.

J1850 is a fairly typical medium-speed, general-purpose data bus using arbitration and a bus topology.

Although J1850 has a companion conformance specification, which is J1699, compliance is not guaranteed by any third-party organization. IC makers are not licensed to make J1850 devices, and the ECUs, of which they become part, do not have to pass a rigorous certification process. Sometimes, the vehicle manufacturers perform some tests at the network level, but generally they rely on their systems suppliers to give them a product that conforms to J1850.

There are several documents, such as J2178 [5, 6–9], that are always used in conjunction with J1850 to make a complete system. More details can be found in Chapter 4.

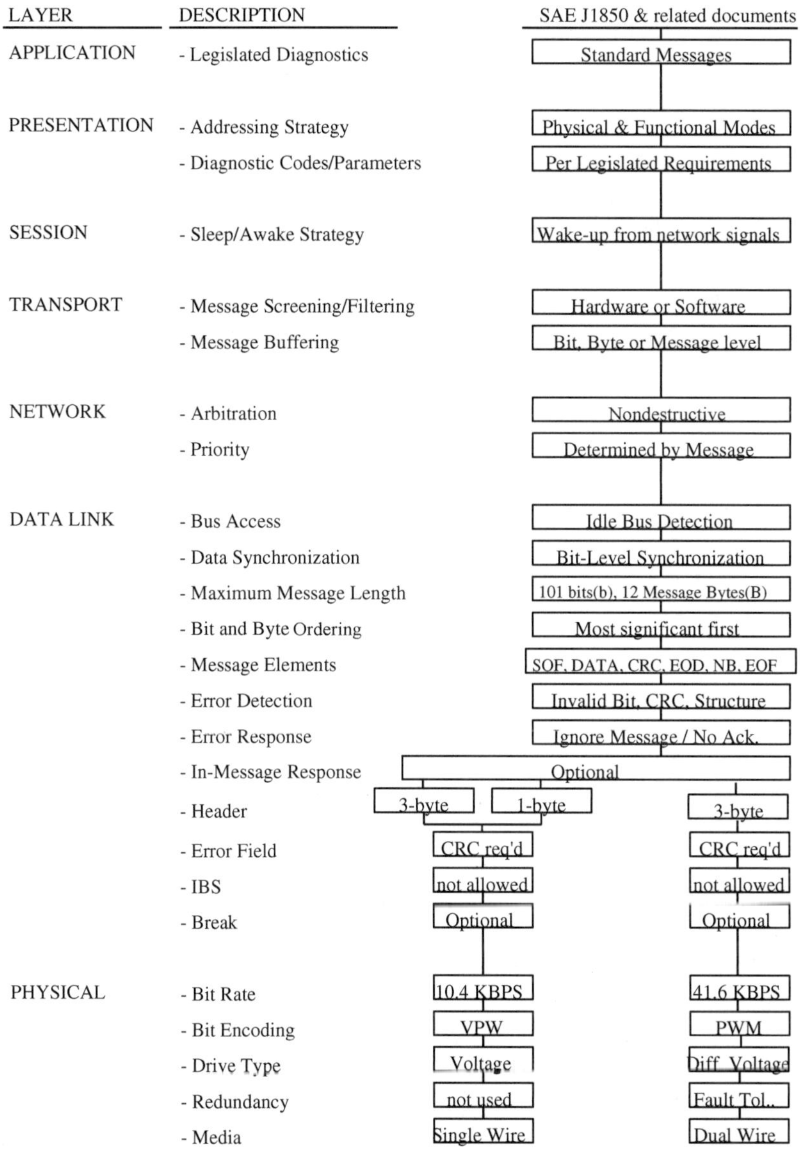

Fig. 14.5 J1850 as applied to a seven-layer model.

Class 2 Introduction

A little bit of background is in order for Class 2. In the mid-1980s, GM was already categorizing its serial bus protocols much like the strategy that the SAE would adopt some time later. The GM classifications were Class 1, Class 2, and Class 3. Class 1 protocols were for high-speed, real-time, critical-control applications. Class 2 encompassed medium-speed, diagnostic, and information-passing designs [10]. Class 3 was the group of low-speed, non-critical, low-function protocols. By 1985, development was underway within the Audio Systems unit of Delco Electronics on a completely new, medium-speed, serial data protocol. Unfortunately, the moniker "Class 2" became the name for this protocol.

Around the same time, SAE began a classification scheme of its own for serial data link protocols. SAE used Class A, Class B, and Class C to describe the types of protocols, mainly by speed. SAE Class A corresponded to GM Class 3, SAE Class B to GM Class 2, and SAE Class C to GM Class 1. Since about 1992, GM and Delco Electronics dropped the Class 1, 2, and 3 designations in favor of the SAE descriptions. This is just as well because the term "Class 2" was now the name of a particular protocol anyway.

In the late 1980s, the California Air Resources Board (CARB) began action to implement a new, more stringent set of emission requirements for all passenger cars sold in California, beginning in model year (MY) 1994. These new regulations would be called On-Board Diagnostics-II, or OBD-II. A key feature of OBD-II would be that it would require a standardized serial data link on each vehicle, regardless of manufacturer, for interfacing an offboard handheld scan tool with the vehicle electronics. This link would receive requests from the scan tool for certain emission data stored in the vehicle, such as the number of misfires per mile and which cylinder(s) are misfiring.

In 1988, CARB began soliciting ideas for the standard data link, which would be described in SAE J1850. GM proposed its new Class 2 protocol, Ford offered a PWM-based solution (SCP), and DaimlerChrysler offered its UART-based DaimlerChrysler collision detection (CCD) bus. CARB turned down the CCD offering but, unfortunately, accepted both the GM and Ford protocols. The data rates were modified for some reason, however. Class 2 was originally at 10 kb/s but was changed at the request of CARB and SAE to

10.4 kb/s (a 96-μs average bit time). Ford's system became the standard corporate protocol (SCP) internal to Ford and runs at 41.6 kb/s.

The direction at GM is to use the Class 2 protocol for all Class B functions, not simply for emissions-related diagnostics. GM's maximum usage for MY96 was one to ten Class 2 nodes with a fleet average of about three nodes per vehicle. This usage is now from three to thirty-two nodes per vehicle since MY00, with an average of about six nodes per application.

Once CARB adopted Class 2 (within J1850) as a standard (see Chapter 4), further development of Class 2 was transferred to Delco Electronics' Powertrain Electronics unit. This was necessitated by the fact that Powertrain would need to have Class 2 in all of its controllers by MY96 to meet OBD-II requirements.

The mandate within GM was to use Class 2 for all general-purpose information passing and diagnostics. The intent was to replace all usage of the UART and E&C buses. By 1990, it was obvious that this was going to be a difficult task.

Most other GM electronic architects were quite happy with their current usage of UART or E&C, even within the Audio Systems group. These Class A buses were mature, proven, and very inexpensive. Getting their products to adopt Class 2 sometimes required a sort of blackmail. This can be illustrated by the story of a new Corvette radio, which needed vehicle speed to incorporate speed-compensated volume. The ECU in the vehicle that contained vehicle speed was the powertrain controller module (PCM). Normally this parameter could be passed to the radio via the E&C or UART bus, but the Powertrain group stipulated that it would be provided only in Class 2 format. This was a key driver in getting at least that one audio program to adopt Class 2 in its product.

The Class 2 bus is both a subset and superset of J1850. It is a subset in that it implements only the one-wire, VPW form of J1850. It is a superset of J1850 because it also specifies several additional features.

One of these additional features is the "4x" or high-speed mode. This is the option to run the Class 2 bus at 41.6 kb/s (still VPW) during certain nonnormal uses, such as flash memory programming.

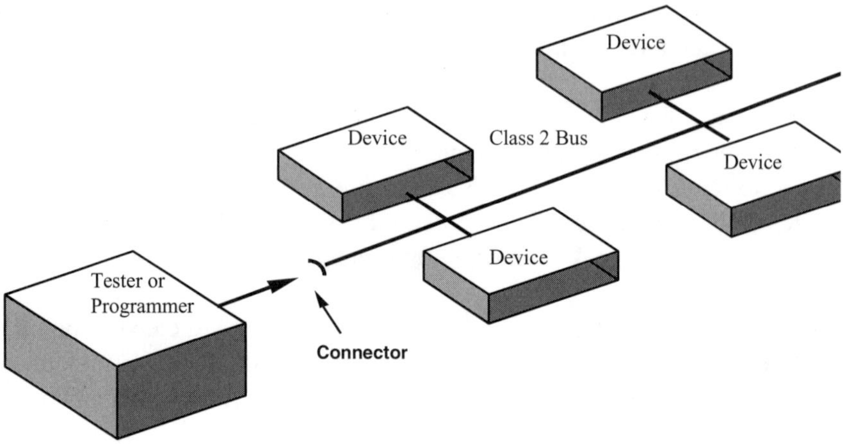

Fig. 14.6 A typical Class 2 network conceptualization.

Another addition is optional unlimited message length. During nonnormal usage, messages longer than 12 bytes may be used. Usually, this "block" mode is combined with the 4x mode.

The Class 2 protocol requires an additional feature that a transmitter must not stop sending information on a byte boundary after the third byte in a message. This retains the correct message length if noise forces a loss of arbitration past the source ID (third byte). Some method must be used to ensure that a message shortened because of noise cannot be taken as correct at a receiver because the final byte matches the CRC for the preceding bytes. The usual method is to have the transmitter automatically transmit two extra "one" bits if arbitration is lost on the eighth bit of a byte.

One other requirement of the Class 2 protocol is that arbitration should be lost anytime the symbol or bit detected on the bus does not match that which was intended to be sent. This means that if noise forces a zero bit to be detected as a one bit, the transmitter should cease contending for the bus.

A companywide database of approved Class 2 messages is supported within GM and Delco Electronics. New messages must be approved by a Class 2 message strategy group that meets once per month.

Finally, Delco Electronics and General Motors do not license or restrict Class 2 technology in any way. The intent has always been that Class 2 should be a world standard and be freely used by IC/module vendors, system suppliers, and vehicle manufacturers.

New ICs or modules, either digital or analog, are not required to be validated or approved by Delphi or General Motors.

The Data Link Controller, Series/Parallel

The data link controller (DLC) circuit was developed by Delco Electronics to implement the Class 2 protocol in a single metal oxide semiconductor (MOS) IC. Initially specified by the Audio Systems (AS) business unit, the DLC was to have an 11-byte transmit FIFO (first-in first-out) (TFIFO) and a 20-byte receive FIFO (RFIFO). A serial peripheral interface (SPI) protocol data interface to the host microprocessor would be provided with interrupt-based servicing assumed. A transceiver would also be built into the IC by fabricating the entire chip with 40-volt high-voltage Complementary Metal Oxide Semiconductor (CMOS) (high-voltage complementary metal oxide semiconductor [HVCMOS]) technology. The part was initially named the DLC-S (data link controller—serial). IC-Delco owns a sophisticated IC tester and was the first group to test DLC breadboards—four-board mock-ups containing hundreds of ICs to simulate the DLC functionality. IC-Delco also did the first testing on the actual ICs because they were available on manufacturing test equipment in FAB III.

The first DLC-S was available for lab evaluation around mid-1988 as the DLC-01. The next iterations of the DLC-S were the DLC-02, DLC-03, DLC-04, and DLC-04A. Each of these early DLC ICs contained only a functional SPI interface to the host and were packaged in 40-pin ceramic-leaded DIP, except for the DL-40A, which was also available in a 52-pin PLCC. Each version had only a 4-byte TFIFO and RFIFO. The DLC-04 and DLC-40A were the first to contain the transceiver and were available in late 1988. Early testing was done by AS, which found that the waveshaping and thus RFI properties were excellent on the DLC-04A. ESD pin protection, however, was not satisfactory and would have to be improved on later versions. Around early 1989, Class 2 and thus the DLC were accepted by SAE as part of the J1850 Serial Data Communications Standard required by the California Air Resources Board on all vehicles sold in California beginning in 1994.

In early 1990, the Automotive Electronics (AE, or Powertrain) business unit assumed responsibility for Class 2 and hence the DLC program. It was determined that a parallel data interface was of paramount importance to the first applications using the DLC. Rather than having both SPI and parallel interfaces available on every DLC, two versions of the DLC were specified: the DLCS and the DLCP. These two ICs would be referred to by a single designation, the DLCS/P IC. AES also would require the ability to do "polled" servicing of the DLCS/P. The DLCS and DLCP use the same die and differ only in a final metal mask step during fabrication that determines if the chip is a DLCS or a DLCP. They do, however, have different part numbers. AES was also doing testing and was the first to interface a DLC with a Motorola microprocessor, the GMP6 (68HC11F1). The first production-like software for the DLCS started in early summer 1990 and was tested on a Corvette electrical system. Two setups were used—a 1991Corvette ECM, CCM, radio head, and instrument panel (IP) on a bench and an actual 1991 Corvette car outfitted with Class 2 communications, instead of UART and E&C.

The first "modern" DLC, the DLCS01, was manufactured by Delco and was widely available in sample quantities in April 1990. The first parallel interface DLC, the DLCP01, was available with the DLCS02 in September 1990. None of them had acceptable waveshaping, ESD, or RFI characteristics. A version of each (DLCS03/DLCP02) to fix a few bugs and add a message length enhancement was available in late August 1991. The first production was built by IC Delco and Motorola in late October 1991 as the DLCS04A/DLCP03A. This was an attempt to optimize EMC, and ESD characteristics added the capability of transmitting an in-message response without CRC. Ultimately, DLCS05/DLCP04 and DLCS05A/DLCP04A versions were needed to fix bugs and make a further attempt to increase the radiated immunity of the transceiver. The final version of the DLC, the DLC08A, combines the serial and parallel CPU interfaces on a single chip. The interface chosen is enabled via a moding pin.

Until the DLCP07, all DLCs were available only in the plastic leaded chip carrier (PLCC) package. This package was chosen to minimize circuit board area and provide surface-mount capability. A small outline (SO) package was subsequently made available as an option for all DLCs.

Pin-outs of the DLCS, DLCP, and DLC08A are shown in Fig. 14.7.

Both Motorola and TI were licensed to produce the DLC. Only Motorola opted to manufacture such a standalone device. Its equivalent to the DLCP is the MC68HC56, the DLCS is the MC68HC57, and the DLC08A is the MC68HC58. These ICs are commercially available to anyone and are priced at less than $2.00 each in quantities greater than one million.

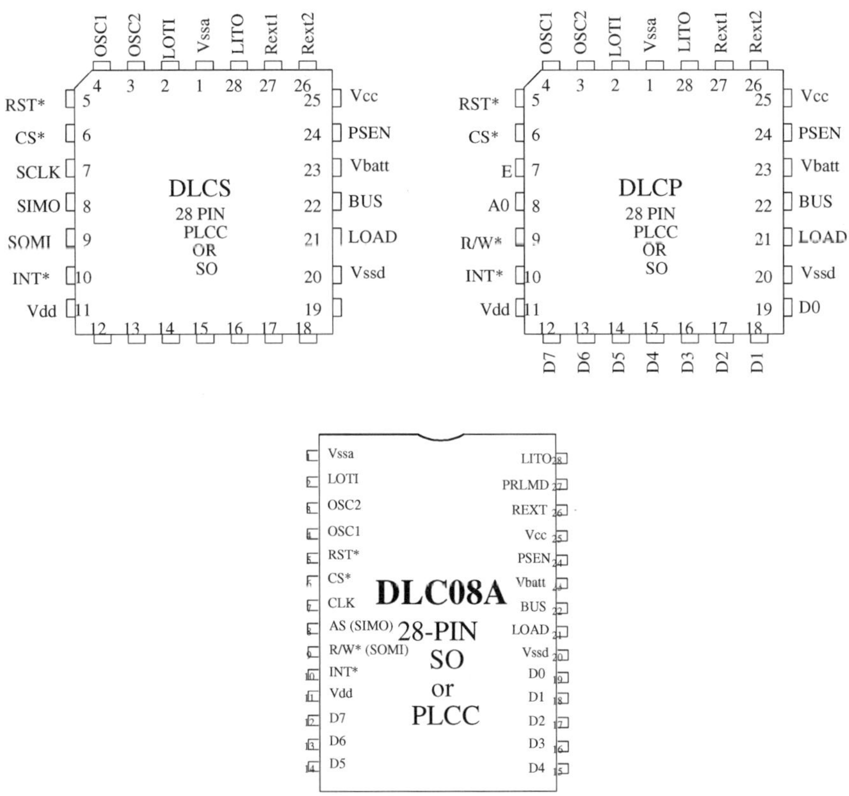

Fig. 14.7 DLC pin-outs.

DaimlerChrysler Implementation of J1850

DaimlerChrysler also uses the one-wire VPW form of J1850 but applies the 1-byte header of J2178 in addition to the 3-byte header. DaimlerChrysler also uses in-frame response (IFR) in some cases with a polarity opposite from that of GM on the first bit of the IFR (the normalization bit [NB], or n/f bit).

This form of J1850, sometimes referred to as a programmable controller interface or "PCI" within DaimlerChrysler, will not be further described here. A Class 2 protocol digital device can be compatible with the DaimlerChrysler implementation if the polarity of the n/f bit is programmable and if the device supports all four types of IFR per J2178.

Summary of One-Wire VPW J1850 Hardware

A list of some commercially available one-wire J1850 ICs is given in Table 14.1.

TABLE 14.1
COMMERCIALLY AVAILABLE ONE-WIRE J1850 ICS

Part	Manufacturer	Package	Type	Module Name	GM or Chrysler
MC68HC58	Motorola	28-PLCC, SO	Digital & Analog	N/A	GM
HC08/12	Motorola	Various	Digital	BDLC	GM & Chrysler
68377	Motorola	Various	Digital	DLCMD2	GM & Chrysler
MC33990	Motorola	8-SO	Analog	N/A	GM & Chrysler
PRISM family	TI	Various	Digital & Analog	CDC & CRD	GM & Chrysler
AU5780/83	Philips	Various	Analog	N/A	Chrysler
C161 family	Infineon	Various	Digital	SDLM	GM & Chrysler
TC1775	Infineon	Various	Digital	SDLM	GM & Chrysler
14056	AMI	14-SO	Digital	JBLIC	Chrysler

A comparison of one-wire, VPW digital hardware forms of J1850 is given in Table 14.2.

A comparison of one-wire, VPW digital plus analog hardware forms of J1850 is given in Table 14.3.

TABLE 14.2
COMPARISON OF ONE-WIRE J1850 DIGITAL HARDWARE PARTITIONS

	Symbol	Byte	Message
Advantages	• Cheap • Small H/W size	• Average price • Average size	• Little code
Disadvantages	• Lots of code	• Average code	• Large H/W size • Expensive
Hardware Examples	• Harris 'Sendec' on '05	• Motorola BDLC on '08 • AMI JBLIC on 14056 • Intel on '196LB • TI on PRISM	• DLCS/DLCP • Motorola MDLC on '05 • Motorola DLCM on '332

TABLE 14.3
COMPARISON OF ONE-WIRE J1850 DIGITAL + ANALOG HARDWARE PARTITIONS

	µ + dlc + idr	µ + idr	µ
Advantages	• Very flexible • Easiest to design/build	• Sort of flexible • Average cost • Average size • Easier to fab µ	• Cheapest (maybe) • Smallest (maybe)
Disadvantages	• Many chips • Most expensive	• Extra chip	• Very inflexible • Hard to fab
J1850 VPW Hardware Examples	• AMI 14056+ Philips AU5780	• DLCS/DLCP • Intel • Moto '332	• Moto HC05Vx • TI PRISM

Standard Corporate Protocol (SCP)

The other major form of J1850 is used by Ford and called standard corporate protocol (SCP). This is the two-wire PWM implementation. Many difficulties were encountered in developing SCP, and it began to be phased out after MY00 in favor of CAN.

Wayne Johnson and Jack Volk of Ford first described SCP in 1986 [11].

Although a transceiver IC was supposed to be made available, none was successful, so the discrete circuit seen in Fig. 14.9 is normally used with a digital IC, such as the LBCC (Fig. 14.8). An earlier part, the HBCC, referred to as Motorola SC370621, was not generally available and is discontinued now. The LBCC is still available from Motorola.

SCP functionality is also provided on the Motorola MC68HC05X1 microcontroller. This is a commercially available IC, as are a pair of standalone ICs from Mitsubishi called the M64100TFP and M64105P.

The SCP transceiver circuit in Fig. 14.9 approximates the bus waveforms in Figs. 14.10 and 14.11.

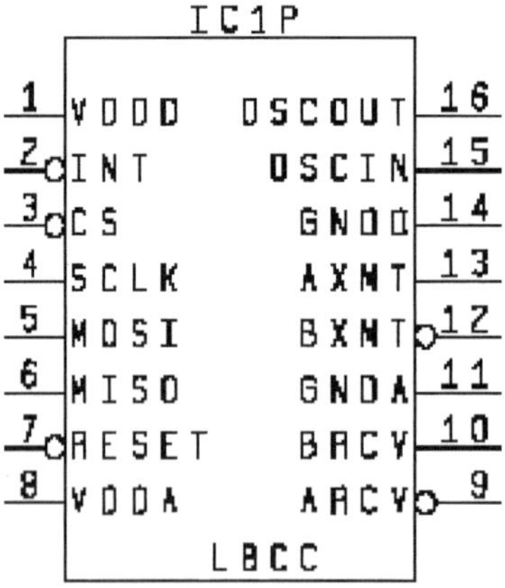

Fig. 14.8 LBCC pin-out.

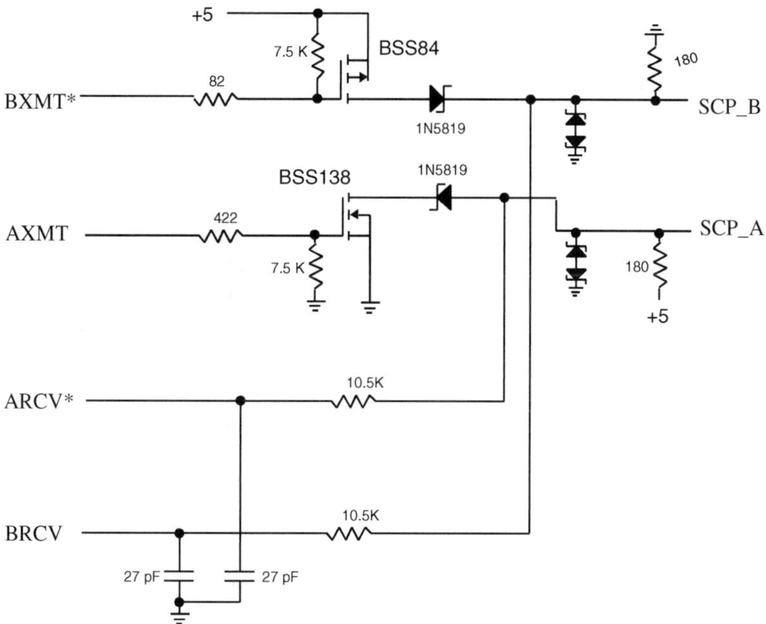

Fig. 14.9 SCP transceiver circuit.

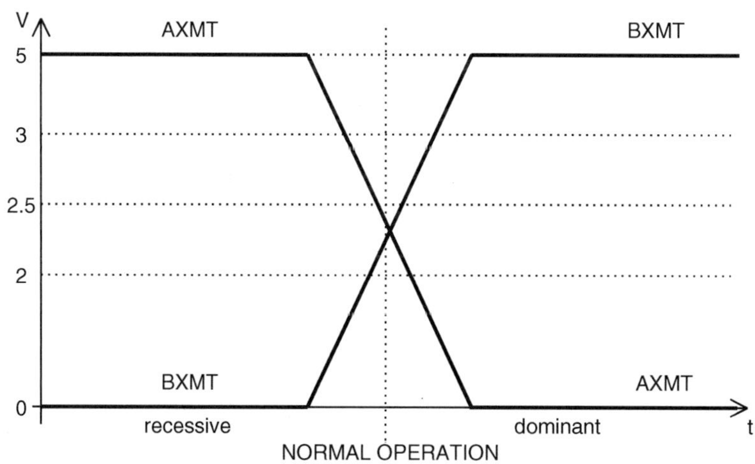

Fig. 14.10 SCP normal waveform.

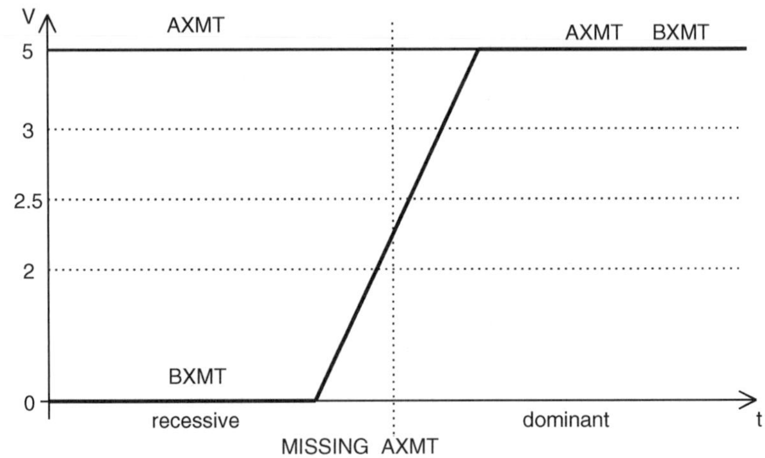

Fig. 14.11 SCP waveform with stuck wire.

Summary of Two-Wire PWM J1850 Hardware

Motorola LBCC

- 16-pin SO (digital only)

Motorola MC68HC05X1

- MC68HC05 CPU
- SAE J1850 PWM serial multiplex interface (SMI)
- 44-pin PLCC package (digital only)

Mitsubishi M64100TFP

- Standalone IC (digital only)
- 44-pin QFP

Mitsubishi M64105P

- Standalone IC (digital only)
- Limited buffering
- 20-pin DIP

A comparison of the two main forms of J1850 is given in Table 14.4.

TABLE 14.4
COMPARISON OF J1850 MAJOR FORMS

	Two-Wire J1850 (PWM)	One-Wire J1850 (VPW)
Advantages	• Faster • Operates with one wire	• Cheaper • Simpler • One-chip solutions • Fewer wires
Disadvantages	• More edges (noisier) • Complicated transceiver • More wires • No one-chip solutions • More expensive	• Slower • No redundancy • Difficult encoding

Single-Wire CAN

Single-wire CAN (SWC) is likely to become an industry standard Class B data link around the MY06 time frame. A new SAE task force has been started (J2411), as well as an ISO group, to specify SWC.

SWC will use the same digital hardware as two-wire CAN but will need a new physical layer because SWC uses only one bus wire. Several new transceiver ICs will be developed to implement SWC. These ICs will be alternatives for two-wire CAN transceivers, such as Philips TJA1050.

Right now, SWC allows as many as 32 nodes on 60 meters of bus. It uses the standard (11-bit) CAN message IDs and is, obviously, a single-wire medium. The data rate is 33.33 kb/s with an optional 83.33-kb/s high-speed mode for memory programming or other nonnormal operations. An additional feature is a sleep mode, whereby nodes can be selectively put in a low-power state and ignore bus traffic. Sleeping nodes are awakened by a special high-voltage message (around battery level).

General Motors is the primary developer of the SWC physical layer and began implementing it in MY02 on certain European vehicles.

Motorola Interconnect Bus

The Motorola interconnect (MI) bus was developed specifically for motor-control applications. It has an interesting message format that is shown in Fig. 14.12.

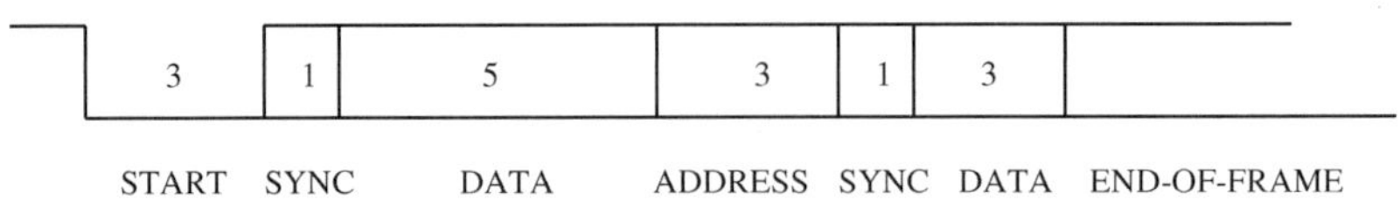

Fig. 14.12 MI bus message format.

The MI bus operates over a single wire at a data rate of 20 kb/s. Three different bit encodings are used. Most of the bit fields in the message are Manchester biphase L. The data field is NRZ, and a 20-KHz square wave signifies the end-of-frame.

A couple of commercially available chips implement the MI bus:

- MC33192DW (restricted sampling)
- Serial stepper motor controller
- MC68HC(7)11P2
- 32K (EP)ROM, 1K RAM, 640 EEPROM, timer, 62 I/O, A/D, PWMs, and so forth

References

[1] Ortis, M., Lacrouts, I., and Herbault, P. "VAN—The Optimized Industrial Solution for Vehicle Multiplexing (Vehicle Area Network)." SAE Paper No. 920223, Society of Automotive Engineers, Warrendale, PA, 1992.

[2] Artur, P. "VAN (Vehicle Area Network) Conformance Methodology." SAE Paper No. 950297, Society of Automotive Engineers, Warrendale, PA, 1995.

[3] ISO 11519-3 Vehicle Controller Network. International Standards Organization.

[4] SAE J1850. Society of Automotive Engineers, Warrendale, PA.

[5] Zachos, M., and Pohlmeyer, A.J. "Message Structure and Strategy to Drive SAE J1850 Networks—An Introduction to SAE J2178." SAE Paper No. 920227, Society of Automotive Engineers, Warrendale, PA, 1992.

[6] SAE J2178/1. Society of Automotive Engineers, Warrendale, PA.

[7] SAE J1278/2. Society of Automotive Engineers, Warrendale, PA.

[8] SAE J2178/3. Society of Automotive Engineers, Warrendale, PA.

[9] SAE J2178/4. Society of Automotive Engineers, Warrendale, PA.

[10] Lupini, C., and Braun, T. "Class 2: An Introduction to Medium Speed Multiplexing." SAE Paper No. 920222, Society of Automotive Engineers, Warrendale, PA, 1992.

[11] Johnson, W., and Volk, J. "A Proposal for a Vehicle Network Protocol Standard." SAE Paper No. 860392, Society of Automotive Engineers, Warrendale, PA, 1986.

Chapter 15

Class C Protocols

Introduction

This chapter discusses some high-speed protocols for automotive applications. Most of the attention centers on CAN, although there are many variations of CAN, some not used for Class C implementations.

Multiplex Standardization

Table 15.1 shows the major divisions of the (more or less) standardized protocols in use throughout the world. The only thing missing is ISO 9141, which is heavily used in Europe and Asia for manufacturer-specific diagnostics.

CAN

CAN is the major automotive serial data protocol in use throughout the world. Unfortunately, there are several variations from which to choose, and more

TABLE 15.1
WORLD MULTIPLEX STANDARDIZATION CHART

	Transfer Rate	
Organization	**< 125 kb/s**	**> 125 kb/s**
ISO (Europe)	CAN ISO 11519 ISO 11992	CAN (ISO 11898)
SAE (U.S.) passenger cars	Class 2/SCP/etc. (J1850)	CAN (SAE J2284)
SAE (U.S.) truck and bus	J1587/1708, J1922	CAN (SAE J1939)
ASIA	none	CAN

are becoming available all the time. Table 15.1 shows the breakdown by region, and Fig. 15.1 shows the outgrowth of the variations from the Bosch CAN 2.0 [1] specification (CAN family tree).

Table 15.2 roughly compares and contrasts the major commercial "flavors" of CAN.

The Bosch CAN specification outlines only the data link layer and part of the physical layer of the OSI model. This is illustrated in Fig. 15.2.

Because the Bosch CAN specification covers only the data link and a small portion of the physical layers, there are many "flavors" of CAN. These protocol flavors fill in many of the gaps from the higher layers to make a more complete structure. For an example, refer to Fig. 15.3.

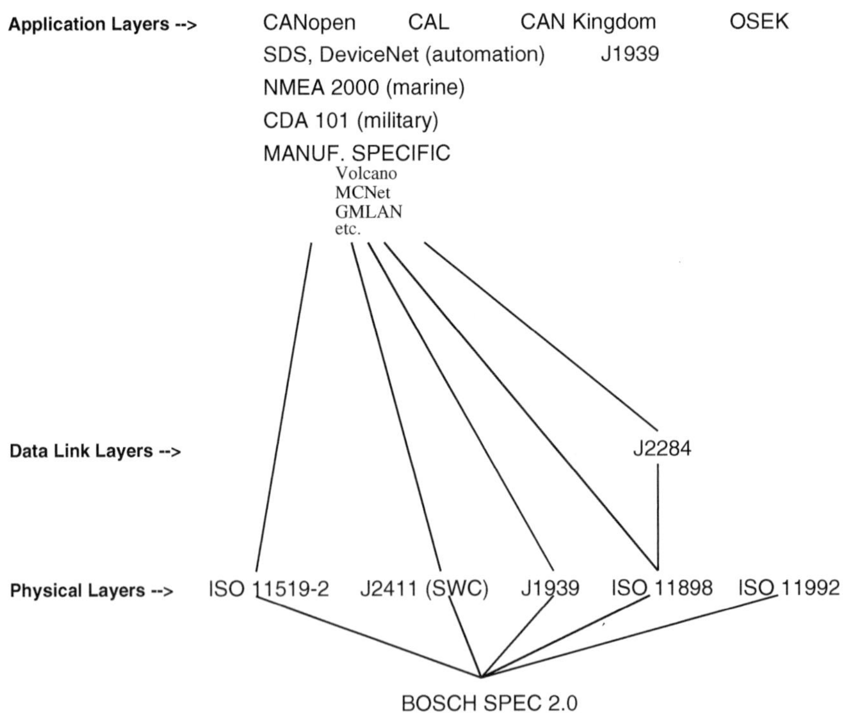

Fig. 15.1 CAN protocol family tree.

TABLE 15.2
COMPARISON OF CAN VARIATIONS

Protocol	Uses	Advantages	Drawbacks
Bosch 2.0	• Class A functions • Class B functions • Class C functions	• Widely used	• Very limited in details • No physical layer • No application layer • No messaging spec. • Not OBD-II compatible
ISO 11898	• Class C functions	• Widely used	• Europe only • 125 kb/s • 1 Mb/s • Not OBD-II compatible
ISO 11519	• Class A functions • Class B functions	• Fairly simple • Easy to do fault-tolerance. • 10 –125 kb/s	• No wide buy-in • Europe only
ISO 11992	• T & B trailer interface	• Unknown	• Unknown
J2284	• Class C functions	• U.S. standard	• One speed (500 kb/s) • Limited message definition
J1939	• Class C functions	• Complete • Good truck standard • Getting wide buy-in • Compat. w/European Agriculture	• Unique msg. strategy • Not good for cars
J2411	• Class A functions	• Inexpensive • One wire • Same CAN digital	• New transceiver • Slow 33.33 kb/s • GM only (for now)

Controller Area Network (CAN) Background

Controller area network (CAN) is a serial data link protocol suitable for real-time, safety-critical, vehicle control applications. It was intended to support distributed, parallel, object-oriented processing. Development by Bosch began in the early 1980s, and Intel became a major partner about 1985. Although

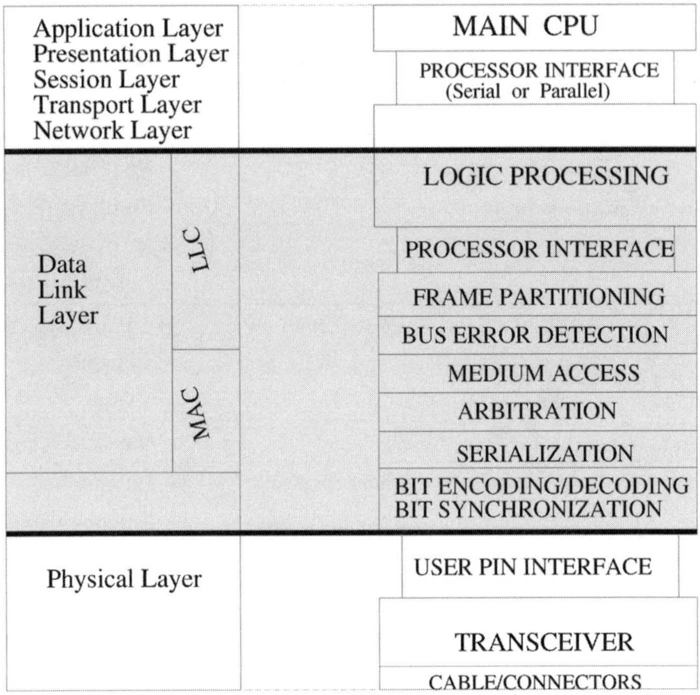

Fig. 15.2 CAN mapping to OSI model.

meant strictly for automotive uses, there are now more nonautomotive nodes than there are nodes in vehicles. The protocol was first described commercially in the United States via SAE J1583 [2]. J1583 was essentially a copy of the Bosch specification but also included an appendix that contained an overview of the Intel 82526 IC. An interesting early paper by the "fathers of CAN" (Kiencke, Dais, and Litschel) can be found in [3]. CAN had to beat out VAN and ABUS, and more recently, Advanced PALMNET, for dominance [4].

The two key features of CAN that make is desirable from a Class C viewpoint are its error handling (15-bit CRC, error counters, error frame, etc.) and its bit synchronization (dynamic sample point).

IC vendors are licensed by Bosch to create digital CAN chips and modules. This does not apply to the physical layer (transceiver) because the CAN

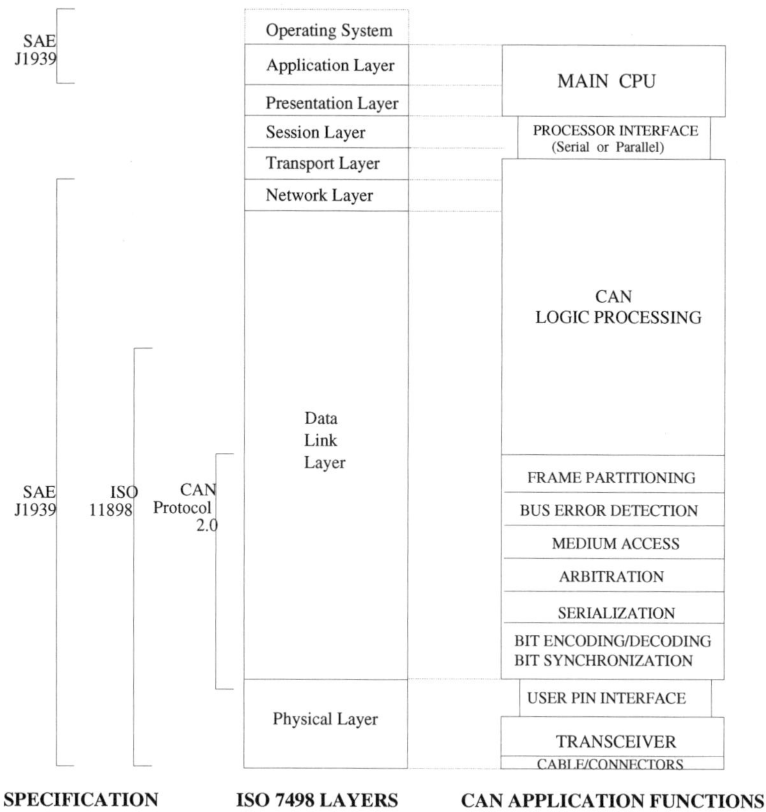

Fig. 15.3 CAN protocol handler layering.

specification does not address (much of) the physical layer. Intel was the first maker of a CAN digital chip. Intel has paid no royalties because it lent such a big hand in the development of CAN. None of the subsequent IC vendors seem to pay royalties either, although there have been some interesting "horse trades" between vendors. For example, witness the Intel 82527 design on a Siemens 80C167C.

When a company is licensed to produce CAN (digital) devices, the company has access to the Bosch verification models. These are test suites either in C or VHSIC Hardware Description Language (VHDL) to help the IC maker check the conformance of their part(s). Use of the models is considered mandatory, but Bosch does not closely monitor the users.

Another organization has come up with CAN conformance tests, the French "Realnet" consortium. A VHDL platform for Realnet is ready for sale by Dassault Electronique. The interesting thing about this is that Dassault will sell their product to anyone, whereas Bosch limits the availability of its models to CAN licensees.

ISO is trying to standardize CAN conformance via an ad hoc task force under ISO/TC22/SC3/WG1. The goal of this task force is to specify an international standard containing CAN conformance tests. ISO believes that this is very important because the CAN patents will expire soon, and anyone may then develop silicon chips without using the Bosch model.

The following IC vendors produce CAN circuitry now:

- AB Automotive
- Fraunhofergesellschaft IMS
- National Semiconductor
- ALCATEL
- MIETEC
- Fujitsu
- NEC
- AMP, Inc.
- Hewlett–Packard
- Okidata
- Analog Devices Inc.
- Hitachi
- Philips
- Atmel TEMIC Semi GmbH
- Intel
- SICAN
- Atmel MHS S.A.
- Infineon Technologies
- ST Microelectronics
- Aurelia Microelettronica
- ITT
- Swedish Space Corp.
- Dallas Semiconductor
- LSI

- Texas Instruments
- DENSO Corp.
- Microchip
- Toshiba
- DESCON GmbH
- Mentor Graphics
- Vishay Siliconix
- EMS Technologies Canada
- Mitsubishi
- ZAM
- FH–Wolfenbuttel
- Motorola

The first application of CAN was by Mercedes on its S-class models in the 1991 model year. This vehicle line had five CAN nodes and used a 500-kb/s bit rate.

A time line of CAN progress is given in Table 15.3. Of special interest is the fact that the Bosch patents on CAN allegedly began expiring in 2001. This could mean the advent of lower-priced chip sets from the increased competition.

Approximately 9 million CAN ICs were sold from 1989 through 1994. About 14 million were sold during 1996, and 40 million in 1999. It is forecast that there will have been more than 300 million nodes in use around the world at the end of 2003 [5].

As mentioned previously, CAN is now used in many places other than vehicles. The following are some of them:

- Elevators
- Escalators
- Agricultural machinery
- Medical systems
- Textile production
- Laser printers
- Home appliances
- Vending machines

- Robots, numeric machine tools
- Office control systems
- Trains (including model railways!)
- Ships
- Car washes
- Planes (some DC-10s)

TABLE 15.3
CAN TIME LINE

1983	Start of development
1984	
1985	Specification 1.0
1986	
1987	Introduction of first chip—82526
1988	
1989	First microcontroller with CAN—MC68HC05X4
1990	
1991	Specification 2.0: First production CAN auto application
1992	
1993	
1994	Completion of ISO standardization; J1939 usage
1995	
1996	
1997	
1998	
1999	
2000	CAN acceptable for E-OBD usage
2001	CAN patents expire (?)
2002	
2003	
2004	Widespread use of CAN in U.S. passenger cars
2005	

The current CAN specification is on revision 2.0, which dates from 1991. This document is actually composed of two parts, 2.0 A and 2.0 B. CAN 2.0 A is sometimes referred to as BCAN , BasicCAN, or CAN 1 (ISO). CAN 2.0 B has been called FCAN, FullCAN, E-CAN, Extended CAN, or CAN 2 (ISO).

CAN 2.0 A is essentially the same as the previous CAN specification version 1.2. CAN 2.0 B is identical to CAN 2.0 A, except for the required message header field that is 29 bits instead of 11 bits.

General CAN Design Issues

For application information, refer to the manufacturer's data sheets for the parts you are using. However, the next few sections will give some tips on using any digital and analog components in a node circuit design. These "rules of thumb" assume that the reader is familiar with the CAN protocol.

For a very thorough treatise on engineering with CAN, refer to the excellent book [6] by Dr. Wolfhard Lawrenz, who was involved in the initial development of the CAN protocol.

Identifier Length

CAN 2.0 A (11-bit) identifiers are generally used. Very few industry applications use 29-bit (extended) identifiers, and initially it was thought that there was no good reason for an application to use them. However, it is being discovered that most CAN application layers "pack data" or put multiple parameters into a single CAN message. This means a smaller message set and lower bus use that can allow a larger message header. J1939 uses 29-bit headers, and GMLAN version 3.0 will also use them.

Clocking

Many of the flavors of CAN used today specify a tolerance on the bus data rate. Typically this tolerance is ±0.5% or less. This means that a crystal will be necessary to provide the clocking for the CAN bus circuitry instead of a ceramic resonator or RC network.

In most cases, an "integral" frequency clock source is necessary to get the required CAN bus data rate. For example, one will not achieve 500 kb/s with a 16.667 MHz local clock.

Tolerating clock mismatches between the nodes requires a bit construction different from that for tolerating propagation delays [7]. For a more in-depth analysis of the issues of oscillator tolerance versus propagation delay, refer to [7, 8].

Sampling

Because of the typical >125 kb/s CAN data rates used, single sampling should be done by all nodes. This is also suggested for 40 to 125 kb/s.

Sample Point

Crystal clock sources must be used at all nodes, so mismatch between nodes is not a large concern. Bus length, however, necessitates that the sample point be relatively late into the bit time. Therefore, the sample point should be programmed to fall between 70 and 80% into the bit, including the synchronization segment (sync_seg).

Resynchronization Jump Width

The resynchronization jump width should be programmed for 10 to 25% of the bit time, including the synchronization segment.

Resynchronization Point

Resynchronization should be done only on recessive to dominant edges to ignore the capacitive loading effects on the dominant to recessive bus edges.

Number of Quanta per Bit

The programmer should strive to use 8 to 25 time quanta per bit. Typical numbers are 8 or 16.

Size of TSEG1

The usual number of quanta is five to eighteen for time segment 1 (TSEG1).

Size of TSEG2

The usual number of quanta for time segment 2 (TSEG2) is two to eight.

Size of SJW

The usual number of quanta for the synchronization jump width (SJW) is one or two. There is usually no reason for SJW not to be equal to TSEG2. The CAN controller will use as much of the SJW as necessary to resynchronize.

CAN Bus Structure

A CAN bus is normally a two-wire, single-channel system terminated at the ends, as shown in Fig. 15.4 (ISO 11898 example):

The typical medium used for CAN buses is shielded twisted pair (STP) or unshielded twisted pair (UTP) that has a characteristic impedance of 120 Ω. The transceiver ICs used to drive the CAN bus are designed to transmit and receive signals with the resulting differential bus resistance.

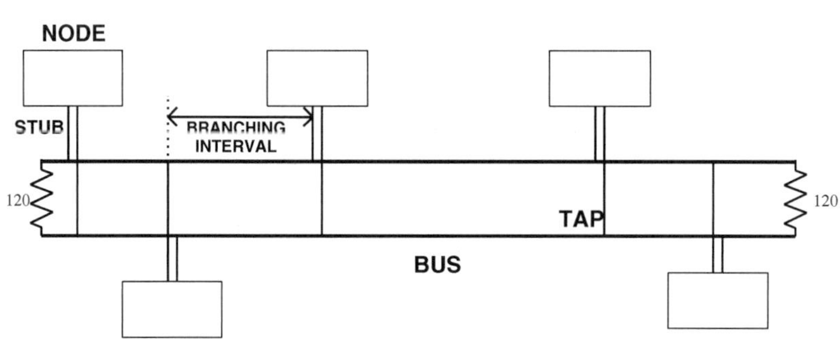

Fig. 15.4 Typical CAN bus structure.

Provision can be sometimes made to allow a third, and even fourth, termination if a long stub (drop) length is necessary. These long stubs may or may not need to be terminated. Transmission theory indicates that each stub should be terminated with the characteristic medium's impedance (120 Ω in this case). Reflections may or may not be a problem and must be checked once the system is built. It may be that terminations on the stubs are not needed. The network provides immunity to reflections via receiver hysteresis and the synchronization rules of CAN.

CAN Termination Strategies

Ideally, *n*odes that do not contain one of the termination resistors should have a short stub length. As a rule of thumb, each unterminated node (stub) should be no further from the main CAN bus trunk than

$$L_u < t_{PROPSEG}/(50 \times t_P)$$

where

L_u	=	maximum stub length for any unterminated node
$t_{PROPSEG}$	=	propagative time (length of TSEG1 minus SJW)
t_P	=	medium propagative delay (assumed to be 5 ns/m)

As an example, putting in the parameters that will be used by a typical system gives a maximum stub length of $L_u < 1.5$ m.

Another rule is that the sum of all stub lengths must be less than

$$L_{u,tot} < t_{PROPSEG}/(10 \times t_P) \text{ for all unterminated nodes}$$

where

$L_{u,tot}$	=	sum of stub lengths for all unterminated nodes
$t_{PROPSEG}$	=	propagative time (length of TSEG1 minus SJW)
t_P	=	medium propagative delay (assumed to be 5 ns/m)

Putting in the parameters that will be used by a typical system gives a maximum total stub length of $L_{u,tot} < 7.5$ m.

CAN Bus Medium

Usually, twisted pair is used for the bus. Purchasing this material from a vendor that sells preformed twisted pair is highly recommended. It is typically an 18 or 20 gauge pair, machine-twisted at a rate of about 30 turns per meter. The pair is jacketed in an insulating sleeve, so that it may not even look like a pair of wires from the outside. It usually looks a lot like phone cord.

For bus length, CAN recommends 40 meters as the maximum distance between any two nodes. However, this is highly dependent on bit rate, clock tolerance at each node, and so forth. Reference [9] finds a 20-meter maximum for 1 Mb/s. Another treatise by the author of this book [7] takes a more detailed look at the trade-offs among bus length, propagation delays, clock tolerance, and so forth.

CAN Transceiver Bus Waveform

Figure 15.5 shows a CAN bus waveform using a Philips 82C250 under severe wiring conditions. This signal is received properly if the bit timing initialization is correct.

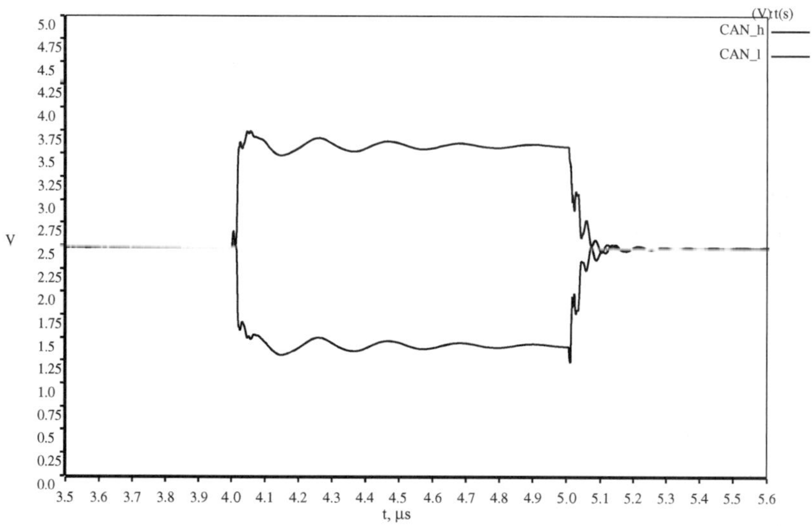

Fig. 15.5 Noisy CAN bus waveform.

CAN Error Recovery

The approximate amounts of time from an error to the next possible message for the error frame scheme that CAN uses, and at 1 Mb/s, are shown in Table 15.4.

SAE J1939

SAE J1939 [10] is a family of recommended practices for U.S. Class C truck and bus (T&B) applications. J1939 is based on CAN 2.0 B (29-bit identifiers) and has various options for physical layer, data link layer, and so forth. It was first used in production during the 1994 model year [11].

Documents associated with J1939 generally have a two-digit suffix, for example, J1939xy. The first digit indicates the corresponding OSI layer number 1–7,

TABLE 15.4
CAN ERROR RECOVERY TIMES

CAN Bus Data Rate = 1 Mb/s		
	Time from Error to Start of Frame (SOF) (μs)	
Scenario	Min.	Max.
Bit error	18.0	150.0
Stuffed error	23.0	140.0
CRC error	54.0	143.0
Form error	52.0	150.0
Acknowledge error	53.0	142.0
Overload error	14.0	46.0
Overload form error	15.0	66.0
Inconsistent overload errors	23.0	173.0
Medium consecutive errors (n = 3)	19.0	190.0
Medium successive errors (n = 3)	—	450.0
Transmitter failure	—	2400.0
Receiver failure	—	2250.0

plus 0 indicating general, and 8 for network management. The second digit has been defined for industry class; 0 indicates index, 1 is for global, and 2 for North America agriculture.

The primary documents that exist today are the following (by layer) [12–19]:

- J1939 (Introduction)
- J1939/81J1939/82
- J1939/71J1939/72J1939/73
- No level 6 documents
- No level 5 documents
- No level 4 documents
- J1939/31
- J1939/21
- J1939/11J1939/12J1939/13J1939/14J1939/15
- J1939/01J1939/02

The entire J1939 set is a very complete recommended practice. The following are some of the things that are specified:

- Bus levels
- Bus levels during arbitration
- Common mode bus voltage range
- Terminating resistor
- Internal resistance
- Differential internal resistance
- Internal capacitance
- Differential internal capacitance
- Bit time
- Internal delay time
- Electrical specifications, including dc, ac, ESD, and so forth
- Topology (bus)
- Connector requirements
- Conformance tests
- How to handle bus failures

The top-layer document is simply J1939 and is an introduction and overview of the protocol. There are also several general-purpose J1939/0y documents:

- J1939/01 specifies the T&B control and communications network
- J1939/02 specifies the off-road machinery control and communications network

The following are variations in the physical layer specifications:

- J1939/11 specifies the 250-kb/s shielded twisted pair
- J1939/12 specifies the 250-kb/s twisted quad
- J1939/13 specifies the 250-kb/s offboard diagnostic connector
- J1939/14 specifies the 1-Mb/s fiber-optic physical layer
- J1939/15 specifies the 250-kb/s unshielded twisted pair

The following are variations in the data link layer specification:

- J1939/21 specifies the data link layer

The following are variations in the network layer specification:

- J1939/31 specifies the network layer

The following are variations in the application layer specification:

- J1939/71 specifies the general application layer
- J1939/72 specifies the virtual terminal interface
- J1939/73 specifies the diagnostics

The following are variations in network management specification:

- J1939/81 specifies the network management protocol
- J1939/82 specifies compliance

For more background and some specific application information, see [20–25].

ISO 11783

ISO 11783 is a close relative of J1939. It is practically a European version of J1939 for its agricultural industry. A comparison of the documentation is shown in Table 15.5.

ISO 11992

ISO 11992 is a close cousin of J1939 [26]. This standard was developed by ISO/TC22/SC3/WG1/TF4. ISO 11992 specifies electrical connections between towing and towed vehicles, including the interchange of digital information via a CAN bus. It is mainly a new physical layer that is more suitable for the harsh environment of T&B applications. ISO 11992 also specifies an application layer for braking and nonbraking equipment. The purpose of this is to facilitate interoperability for trucks and trailers from different manufacturers of braking and other equipment in the trailers via a CAN bus.

TABLE 15.5
RELATIONSHIP BETWEEN J1939 AND ISO 11783 DOCUMENTS

J1939 Component	ISO/DIS 11783 Component
J1939/01 & J1939/02	Part 1: Application Documents
J1939/12	Part 2: Physical Layer (Unshielded)
J1939/21 (29-bit)	Part 3: Data Link Layer
J1939/31	Part 4: Network Layer
J1939/4x	N/A (Transport Layer)
J1939/5x	N/A (Session Layer)
J1939/6x	N/A (Presentation Layer)
J1939/81	Part 5: Network Management
J1939/72	Part 6: Virtual Terminal Applications Layer
Not yet included	Part 7: Basic Messages Applications
Not yet included	Part 8: Implement Messages Applications
J1939/71	Part 9: Powertrain Applications Layer

SLIO CAN

Serial link input/output (SLIO) is a concept for applying the CAN protocol without a host microprocessor. Although it is has not been used in an actual vehicle application, SLIO was hailed as a future solution for smart sensor/actuator buses [27]. SLIO is not likely to see extended use because the LIN protocol is currently available.

GM Token Slot

The GM Research Labs developed the GM "token–slot" (TS) protocol in the early 1980s as a potential future Class C standard. A custom application-specific integrated circuit (ASIC) was developed for evaluation. An SAE Information Report [28] describes the details of the protocol.

As many as 256 data bytes could be included in a TS message data frame. In addition to data frames, there are also token frames. The topology is a token ring, but the medium is not specified. Probably, shielded twisted pair or fiber optics would be used to support the high speed (1–2 Mb/s) of the link. Frame encoding is via NRZ-5.

There was never any production usage of TS.

References

[1] Controller Area Network 2.0. Robert Bosch GmbH.

[2] SAE J1588. Controller Area Network (CAN), An In-Vehicle Serial Communication Protocol. Society of Automotive Engineers, Warrendale, PA.

[3] Kienke, U., Dais, S., and Litschel, M. "Automotive Serial Controller Area Network." SAE Paper No. 860391, Society of Automotive Engineers, Warrendale, PA, 1986.

[4] Lee, S., Nam, S., Yun, J., and Kim, K. "Performance Evaluation of Multiplexing Protocols." SAE Paper No. 981105, Society of Automotive Engineers, Warrendale, PA, 1998.

[5] CAN in Automation (CiA) Newsletter, 10/95.

[6] Lawrenz, W. "CAN Systems Engineering." Springer, New York, 1997.

[7] Lupini, C. "Propagation Delay and Oscillator Drift Analysis for CAN Networks." SAE Paper No. 940362, Society of Automotive Engineers, Warrendale, PA, 1994.

[8] Dietmayer, K., and Overberg, K. "CAN Bit Timing Requirements." SAE Paper No. 970295, Society of Automotive Engineers, Warrendale, PA, 1997.

[9] Tanaka, Suzuki, et al. "High-Reliability Physical Layer for In-Vehicle High-Speed Local Area Network." SAE Paper No. 910464, Society of Automotive Engineers, Warrendale, PA, 1991.

[10] SAE J1939, Introduction and Appendices. Society of Automotive Engineers, Warrendale, PA.

[11] Stepper, M. "J1939 Serial Vehicle Network Explanation and Tutorial." SAE Paper No. 972757, Society of Automotive Engineers, Warrendale, PA, 1997.

[12] SAE J1939/11, Physical Layer—250 kb/s STP. Society of Automotive Engineers, Warrendale, PA.

[13] SAE J1939/12, Physical Layer—250 kb/s Quad. Society of Automotive Engineers, Warrendale, PA.

[14] SAE J1939/21, Data Link Layer. Society of Automotive Engineers, Warrendale, PA.

[15] SAE J1939/31, Network Layer. Society of Automotive Engineers, Warrendale, PA.

[16] SAE J1939/71, Vehicle Application Layer. Society of Automotive Engineers, Warrendale, PA.

[17] SAE J1939/72, Virtual Terminal. Society of Automotive Engineers, Warrendale, PA.

[18] SAEJ1939/73, Application Layer-Diagnostics. Society of Automotive Engineers, Warrendale, PA.

[19] SAE J1939/81, Network Management Protocol. Society of Automotive Engineers, Warrendale, PA.

[20] Zachos, M., and Stone, M. "Class C Communications Protocol Proposal for Off-Road Vehicles." SAE Paper No. 930007, Society of Automotive Engineers, Warrendale, PA, 1993.

[21] Stepper, M. "J1939 High Speed Communications, The Next Generation Network for Heavy Duty Vehicles." SAE Paper No. 931809, Society of Automotive Engineers, Warrendale, PA, 1993.

[22] Buehring, P. "CAN Controller Architecture Optimized for SAE-J1939 Applications." SAE Paper No. 940361. Society of Automotive Engineers, Warrendale, PA, 1994.

[23] Stone, M. "High Speed Networking in Construction and Agricultural Equipment." SAE Paper No. 941662, Society of Automotive Engineers, Warrendale, PA, 1994.

[24] Zachos, M., Buczko, G., Dieball, K., and Stone, M. "Development Tools and SAE J1939 Networks on Tractors." SAE Paper No. 950043, Society of Automotive Engineers, Warrendale, PA, 1995.

[25] Neilsen, M., Tandanga, L., and Stone, M. " Macro Replacement Algorithms for J1939/72 Virtual Terminals." SAE Paper No. 952577, Society of Automotive Engineers, Warrendale, PA, 1995.

[26] ISO 11992 CAN for Tractor-Trailer Interfaces. International Standards Organization.

[27] McLaughlin, R., and Khoh, S. "A SLIO (Serial Link Input/Output) CAN Implementation." SAE Paper No. 941661, Society of Automotive Engineers, Warrendale, PA, 1994.

[28] SAE J2106, Token Slot Network for Automotive Control. Society of Automotive Engineers, Warrendale, PA.

Addtional CAN References:

[29] Gupta, S. "CAN Facilitates In-Vehicle Networking." SAE Paper No. 900695, Society of Automotive Engineers, Warrendale, PA, 1990.

[30] Alverenga, C. "Multiplexing in Automobiles—An Application Example of the CAN Protocol." SAE Paper No. 921504, Society of Automotive Engineers, Warrendale, PA, 1992.

[31] Mathony, H., Kaiser, K., and Unruh, J. "Network Architecture for CAN." SAE Paper No. 930004, Society of Automotive Engineers, Warrendale, PA, 1993.

[32] Szydlowski, C. "A Gateway for CAN Specification 2.0 Non-Passive Devices." SAE Paper No. 930005, Society of Automotive Engineers, Warrendale, PA, 1993.

[33] Szydlowski, C., and Patel, M. "Generic FMEA for Stand-Alone CAN Devices." SAE Paper No. 930006, Society of Automotive Engineers, Warrendale, PA, 1993.

[34] Lawrenz, W. "Network Application Layer." SAE Paper No. 940141, Society of Automotive Engineers, Warrendale, PA, 1994.

[35] Szydlowski, C. "Tradeoffs Between Stand-Alone and Integrated CAN Peripherals." SAE Paper No. 941655, Society of Automotive Engineers, Warrendale, PA, 1994.

[36] SAE J2284, High Speed CAN (HSC) for Passenger Vehicle Applications. Society of Automotive Engineers, Warrendale, PA,

[37] Kato, M., Matsuda, Y., Himono, Y., Nakada, T., and Arima, Y. "Evaluation of Latency Time for Class C High Speed LAN Protocols." SAE Paper No. 940363, Society of Automotive Engineers, Warrendale, PA, 1994.

[38] SAE J2056/1, Class C Application Requirement Considerations. Society of Automotive Engineers, Warrendale, PA.

Chapter 16

Diagnostic Protocols, Airbag Protocols, X-by-Wire Protocols, Mobile Multimedia Protocols, and Wireless Protocols

Introduction

Besides Class A, Class B, and Class C, there are several newer categories of in-vehicle protocols. These include diagnostic protocols, distributed airbag protocols, x-by-wire protocols, mobile multimedia protocols, and wireless protocols. There is overlap in usage among these categories, so many protocols may be used in more than one area or class. A rough idea of cost is given relative to a typical Class B interface (such as a 100-kb/s CAN), which is "1x." So an interface that is "3x" costs about 3 times that of a 100-kb/s CAN interface.

Diagnostics

Diagnostics refers to offboard communication to test tools, programming equipment, and so forth. This could be for emission-related or service-related diagnostic purposes. Additional information can be found in Chapter 4. Usage is to satisfy OBD-II, OBD-III, or E-OBD emission diagnostics, or general service use. If for emission diagnostics, it must be a legally acceptable protocol. Protocols used today (or soon) are listed in Table 16.1.

Because this data link is needed only between the engine controller and the offboard connector, a simple approach is sufficient. Most automakers and

TABLE 16.1
SOME PROTOCOLS FOR EMISSION DIAGNOSTICS

Name	User	Usage	Model Years	Comments
ISO 15765-4	Europe	E-OBD	2000+	E-OBD CAN
J1850	GM, Ford, DC	OBD-II	1994+	Not accepted in Europe
ISO 9141-2	Europe	OBD-II, general	1994+	Old OBD-II UART
ISO 14230-4	Many	OBD-II, OBD-III	2000+	Keyword 2000

truck makers are using KW2000 (ISO 14230) already, so this is rapidly becoming the emissions diagnostic standard. In the United States, high-speed CAN will be phased in, beginning in MY04, as the OBD-III emissions test interface. This will be the only legally acceptable protocol by MY07. SAE J2480 has disappeared and has been replaced with ISO 15765. Table 16.2 gives a comparison of the various protocols for emissions diagnostics.

How ISO 15031 and ISO 15765 are used is shown in Table 16.3.

Airbag Protocols

Usage is for airbag systems and is commonly referred to as "safetybus." There may be two, or more, buses such as one for firing and one for sensing. They must support at least 64 nodes consisting of squibs, accelerometers, occupant sensors, seat-belt pretensioners, and so forth. Cost is (hoped to be) 1x to 2x per node. The USCAR "SafetyBus" committee was attempting to standardize on a suitable protocol but degraded into separate, independent camps. The two main ones are Safe-by-Wire and Bosch Siemens Temic (BST). Byteflight is in production in at least one BMW vehicle. The protocols at the forefront are listed in Table 16.4.

Many issues here involve packaging constraints, existing mechanical envelope, legalities, and so forth. The winning protocol may well be a hybrid of several existing proposals. In fact, BST and Safe-by-Wire are actually hybrids of earlier protocols. Now, there is no clear industry direction. A high-level comparison of features is given in Table 16.5. The key attribute of an airbag protocol is that is must react very quickly. Bit rate is not terribly important, but

TABLE 16.2
COMPARISON OF PROTOCOLS FOR EMISSION DIAGNOSTICS

	Bus Name					
FEATURE	**ISO 15765**	**J1850 ISO 11519-4**			**ISO/DIS 9141 ISO/DIS 9141-2**	**KEYWORD XX**
Affiliation	ISO	GM	Ford	Daimler-Chrysler	World	Various
Application	Emissions diagnostics	General and diagnostics	General and diagnostics	General and diagnostics	Diagnostics only	Diagnostics
Transmission media	Twisted pair	Single wire	Twisted pair	Single wire	Single wire	One-wire
Bit encoding	NRZ	VPW MSb first	PWM MSb first	VPW MSb first	NRZ (strt, 7D, P, stop) LSb first	NRZ
Media access	Tester/Slave	Contention	Contention	Contention	Tester/Slave	Master/ Slave
Error detection	CRC	CRC	CRC	CRC	Parity (odd)	x-bit CS
Header length		32 bits	32 bits	8 bits	Not specified	16 bits
Data field length		0–8 bytes	0–8 bytes	0–10 bytes	Not specified	0–85 bytes
Message overhead		33.3%	33.3%	8.3%	Variable	Variable
In-message response		Optional Normally no	Optional Normally yes	Optional Normally yes	No	No
Bit rate		10.4 kb/s	41.6 kb/s	10.4 kb/s	<10.4 kb/s	5 b/s–10.4 kb/s
Maximum bus length		35 meters (5 meters for scan tool)	35 meters (5 meters for scan tool)	35 meters (5 meters for scan tool)	Limited by total impedance to ground	Not specified
Maximum nodes		32	32	32	Limited by total impedance to ground	10
μ Needed?		Yes	Yes	Yes	Yes	Yes
Sleep/Wake-up		Yes	No	No	No	No
H/W Avail?		Yes	Yes	Yes	Yes	Yes
Cost		Low	Low	Low	Low	Low

TABLE 16.3
APPLICATION OF ISO 15031 AND 15765

OSI Layer	Vehicle Diagnostics	Legislated OBD
Diagnostic application	User defined	ISO 15031-5
Application layer	ISO 15765-3	ISO 15031-5
Presentation layer	N/A	N/A
Session layer	N/A	N/A
Transport layer	N/A	N/A
Network layer	ISO 15765-2	ISO 15765-4
Data link layer	ISO 11898	ISO 15765-4
Physical layer	User defined	ISO 15765-4

TABLE 16.4
CURRENT SAFETYBUS PROTOCOLS

Name	User	Years	Comments
Safe-by-Wire	TBD	2002+	Delphi-TRW-Philips-Autoliv-SDI
BST	TBD	2002+	Bosch-Siemens-Temic
DSI	TBD	2002+	Motorola/AMP
Byteflight	BMW	2002+	"ISIS," SI

minimal latency is paramount. So, a protocol, such as CAN, would not suffice, mainly because of the possibility of message collisions.

X-by-Wire Protocols

Usage is for brake-by-wire, throttle-by-wire, steer-by-wire, and other applications. The bit rate is between 1 Mb/s and 10 Mb/s. Fiber optics will be necessary because of high speed. The utmost in reliability, performance, and real-time capability is required. Cost is around 15x+ per node. Some possible candidate protocols are given in Table 16.6. Note the website addresses where more information is available.

Time-triggered protocol (TTP) has its "foot in the door," but work is underway to see if CAN is capable of doing the job (i.e., time-triggered controller area network [TTCAN]). A major issue is how much fault tolerance is really required. Any scheme will require dual bus interfaces, dual microprocessors, bus watchdogs, timers, and so forth. Cost is a big problem. The level of fault tolerance needed requires a lot of silicon and software, which, of course, is expensive. The consortium TTAgroup (www.ttagroup.org) is trying to standardize on a protocol but (like other consortia) will apparently allow a choice of protocols, either TTP/C or FlexRay. Who knows? In a few more years, TTCAN may also be chosen! These protocols are compared in Table 16.7.

Mobile Multimedia Protocols

Usage is for "PC-on-wheels" applications. At least two different networks and protocols may be necessary. These subcategories will be referred to as

TABLE 16.5
COMPARISON OF SAFETYBUS PROTOCOLS

	Bus Name			
Feature	**BST**	**Safe-By-Wire**	**DSI**	**Byteflight**
Affiliation	Bosch-Siemens-Temic	Delphi-Philips-TRW-Autoliv-SDI	Motorola	BMW
Application	Airbag	Airbag	Airbag	Airbag
Transmission mecia	Two-wire	Two-wire	Two-wire	Two-wire or three-wire or optical
Bit encoding	Manchester biphase	Three-level voltage	Three-level voltage	
Media access	Master/ Slave	Master/Slave	Master/ Slave	Master/Slave
Error detection	Odd parity and/or CRC	8-bit CRC	4-bit CRC	16-bit CRC
Header length	Various	Various	Various	Various
Data field length	1 byte	1 byte	1–2 bytes	1 byte
Message overheac				
In-message ACK.	No	No	No	No
Bit rate	31.25 kb/s, 125 kb/s, 250 kb/s	150 kb/s	5–150 kb/s	10 Mb/s
Maximum bus length	TBD	25–40 m	TBD	TBD
Maximum nodes	12 squibs, 62 slaves	64	16	
μ needed?	No	No	No	No
Sleep/Wake-up	No	No	No	No
H/W available?	Yes	Yes	Yes	Yes
Cost	Low	Low	Low	Low

TABLE 16.6
CURRENT DRIVE-BY-WIRE PROTOCOLS

Name	User	Years	Comments
TTP/C	BMW, Audi	2004+	www.tttech.com
TTCAN	TBD	TBD	www.can-cia.de
FlexRay	BMW, DC, GM, Audi	2004+	www.flexray-group.com

low speed and high speed. Beginning with this book, wireless is now in its own category. The necessary bit rate for mobile media applications is between 250 kb/s and 100 Mb/s+.

Low-Speed Mobile Multimedia

Usage is for telematics, diagnostics, and general information passing. See Table 16.8. Cost is around 3x per node. IDB-C, a token-passing form of CAN at 250 kb/s, has fallen out of favor. Most OEMs already have a midspeed bus based on CAN to handle low-end telematics communication functions. There has been recent interest in a lower-cost "high-end" network that can handle digital audio streams but not necessarily video. To that end, the D^2B and MOST developers are working on copper-based solutions. Depending on the EMC performance, 10 Mb/s, 25 Mb/s, or even 50 Mb/s bit rates are being proposed.

High-Speed Mobile Multimedia

Usage is for real-time audio and video streaming. See Tables 16.9 and 16.10. Cost is around 15x to 25x, mainly because of fiber optics. Fiber optics will be necessary due to the high speed required to pass real-time video streams from multiple sources to multiple outputs. This will probably have to be compatible with industry-standard systems such as Connected Car PC, or AutoPC. D^2B has seen the first usage (Mercedes 1999 S-class), but MOST appears to be the top contender at this time. The IDB Forum is leaning toward Firewire via an effort called "IDB-1394." The Automotive Multimedia Interface Collaboration (AMI-C) has set its sights on MOST and Firewire.

TABLE 16.7
COMPARISON OF DRIVE-BY-WIRE PROTOCOLS

	Bus Name			
FEATURE	**TTP**	**Intellibus**	**FlexRay**	**TTCAN**
Affiliation	U-Vienna	Boeing/SAE	Motorola	CiA
Application	Safety control	Safety control	Safety control	Safety control
Transmission media	Not specified	Twisted pair	Twisted pair or fiber	Twisted pair
Bit encoding	Not specified	Manchester biphase	NRZ	NRZ-5
Media access	Isochronous	Master/Slave	Time or priority	Time or contention
Error detection	16-bit CRC	CRC, parity	24-bit CRC	15-bit CRC
Header length	1 byte	16–48 bits	40 bits	11–29 bits
Data field length	16 bytes	0–32 bytes	0–246 bytes	0–8 bytes
Message overhead	18.75%	28–75%	3–100%	9.9–22%
In-message ACK.	Yes	Optional	No	Yes
Bit rate	Not specified	12.5 Mb/s	10 Mb/s	1–2 Mb/s
Maximum bus length	Not specified	30 meters	Not specified	40 meters
Maximum nodes	Not specified	64	Not specified	32
μ Needed?	Yes	No	Yes	Yes
Sleep/Wake-up	No	Yes	Yes	Yes
H/W avail?	No	FPGA	No	Yes
Cost	High	Low	Medium	Low

TABLE 16.8
SOME LOW-SPEED MOBILE MEDIA BUS PROTOCOLS

Name	User	Model Years	Comments
IDB-C	None long term	2002+	250 kb/s CAN; www.idbforum.org
D^2B SmartwireX	TBD	2005+	www.candc.co.uk/candc_company
MOST over copper	TBD	2005+	www.mostcooperation.com

TABLE 16.9
CURRENT HIGH-SPEED MOBILE MEDIA BUS PROTOCOLS

Name	User	Year	Comments
D^2B	Mercedes, Jaguar	1999+	www.candc.co.uk/candc_company
MOST	BMW, GM, DC, Ford, VW, Toyota	2000+	www.mostcooperation.com
IDB-1394	DC	2000+	www.1394ta.org
USB	Clarion	1998+	www.autopc.com
Intellibus	Boeing	2004+	www.intellibusnetworks.com

Wireless Protocols

Wireless protocol usage is somewhat undetermined at this time. It will be necessary (initially) for cell phones and palm PCs. Eventual use may include cameras, pagers, and so forth. The cost target is around 5x per node.

Much of the advertisement attention has been on Bluetooth. However, 802.11 also has its proponents, so many groups and suppliers are studying collocation, so that the two standards can exist near each other. Ultrawideband (UWB) is the wildcard that some think will "blow Bluetooth and 802.11 away." Approved by the Federal Communications Commission (FCC) in February 2002, it is essentially "white noise" communication. Using precise clocking, tiny amounts of information are transported across a very wide range of frequencies at very low power (perhaps 1/10,000th of that of a cell phone). Compared with spread spectrum, which uses a small range of frequencies one at a time, UWB uses a wide range of frequencies all at once. UWB can be used for global position sensing, it works indoors, and it easily penetrates obstructions.

TABLE 16.10
COMPARISON OF MOBILE MEDIA PROTOCOLS

	Bus Name							
Feature	**IDB-C**	**Intellibus**	**MOST**	**D^2B Copper**	**D^2B Optical**	**MML**	**USB**	**IEEE 1394**
Affiliation	SAE	Boeing/SAE	Oasis	C&C	C&C	Delco		IEEE
Application	Aftermarket entertainment	Control and diagnostics	Stream data and control	Stream data and control	Stream data and control	Stream data and control	PC devices	PC devices
Transmission media	Two-wire	Twisted pair	Optical	Twisted pair	Optical	Optical fiber	Shielded twisted pair	Shielded twisted pair
Bit encoding	NRZ	Manchester biphase	Biphase	PWM	Bphase	NRZ	NRZ	NRZ
Media access	Token–slot	Master/Slave	Master/Slave	Master/Slave	Master/Slave	Master/Slave	Contention	Contention
Error detection	15-bit CRC	CRC, parity	CRC	Parity	CRC	Correcting (optional)	CRC	CRC
Header length	11 bits	16–48 bits				1 byte		
Data length	8 bytes	0–32 bytes				1–200+ bytes		
Message overhead	~ 32 Bits	28–75%				5 –10%	25%	25–30%
In-message ACK.	1 ACK bit	Optional	No	No	No	No		
Bit rate	250 kb/s	30 Mb/s	25 Mb/s	29.8 kb/s	12 Mb/s	110 Mb/s	12 Mb/s	98–393 Mb/s
Maximum bus length	TBD	30 Meters	Infinite	150 meters	No limit	10 meters		72 meters
Maximum nodes	16	64	24	50	24	16	127	16
μ Needed?	Yes	No	Yes	Yes	Yes	Yes	Yes	Yes
Sleep/Wake-up	Yes	Yes				Yes	No	No
H/W avail?	No	FPGA	Yes	Yes	Yes	No	Yes	Yes
Cost	Low	Low	High	High	High	High	Medium	Medium

Leading wireless protocols are listed in Table 16.11.

TABLE 16.11
CURRENT WIRELESS MOBILE MEDIA BUS PROTOCOLS

Name	User	Model Years	Comments
Bluetooth	TBD	2005+	www.bluetooth.com
IEEE 802.11	TBD	TBD	www.ieee802.org/11
UWB	TBD	TBD	www.uwb.org

Bluetooth

Bluetooth is a new technology standard for wireless communication between devices. It offers a short-range radio link with low-power and low-cost modules. It is an open specification aiming for wide use and compatibility. It was invented in 1994 as a radio link between mobile phones and was named after Viking king Harald Blaatand (940–981), the king of Denmark. There are already several production applications of Bluetooth in automobiles. These are mainly for hands-free operation of Bluetooth-enabled cell phones. In the near future, additional personal computing devices such as PDAs should be accommodated.

IEEE 802.11

IEEE Local Area Network Standards:

- 802.0 SEC (Sponsor Executive Committee)
- 802.1 High Level Interface (HILI) Working Group
- 802.2 Logical Link Control (LLC) Working Group
- 802.3 CSMA/CD Working Group
- 802.4 Token Bus Working Group
- 802.5 Token Ring Working Group
- 802.6 Metropolitan Area Network (MAN) Working Group
- 802.7 Broadband Technical Advisory Group (BBTAG)
- 802.8 Fiber Optics Technical Advisory Group (FOTAG)

- 802.9 Integrated Services LAN (ISLAN) Working Group
- 802.10 Standard for Interoperable LAN Security (SILS) Working Group
- 802.11 Wireless LAN (WLAN) Working Group
- 802.12 Demand Priority Working Group
- 802.14 Cable-TV Based Broadband Communication Network Working Group
- 802.15 Wireless Personal Area Network (WPAN) Working Group
- 802.16 Broadband Wireless Access (BBWA) Working Group
- RPRSG Resilient Packet Ring Study Group (RPRSG)

IEEE 802.11 belongs to the Wireless Local Area Networks (WLAN) Group. The specifications include a Medium Access Control (MAC) layer and currently include five physical (PHY) layers:

- Frequency hopping spread spectrum
- Direct sequence spread spectrum
- Infrared
- 11 Mb/s at 2.4 GHz
- 54 Mb/s at 5 GHz

Ultrawideband

Ultrawideband (UWB) technology uses very short (nanosecond) bursts of energy spread over thousands of megahertz. Potential applications include communications—some are potentially synergistic with global positioning system (GPS). But UWB emissions may cross restricted spectrum bands used by GPS and cause interference. Ultrawideband (UWB) describes a number of radio/radar signal types such as impulse, carrier-free, baseband, time domain, and large relative bandwidth. Each of these types of electromagnetic energy propagation can be characterized by widely varying signal structures and temporal variations.

The FCC proposed definition is as follows: any device where the fractional bandwidth is greater than 0.25 or occupies 1.5 GHz or more of spectrum. A signal centered at 4 GHz would be considered UWB if its bandwidth were 1 GHz or more. A signal at 1 GHz would qualify as UWB for bandwidths of 250 MHz or more. UWB can be used by anyone, anytime, anywhere. The FCC approved it on February 14, 2002.

Here is a quick summary of UWB attributes:

- Unlicensed
- Series of short electrical pulses
- Exists in a range of frequencies simultaneously
- A "blast of electrical noise"
- The pulses beat out a code
- The receiver has to know "when to listen"
- Very secure
- The basic concept dates back to the 1960s

Ultrawideband is considered a successor to spread-spectrum (SS) techniques, which use a small range of frequencies, one at a time. Ultrawideband uses a wide range of frequencies all at once. There is a huge potential data-carrying capacity with ultralow power (perhaps 1/10,000th of that of a cell phone). To conventional receivers, UWB communications simply resemble white noise.

Chapter 17

Data Link Usage

Introduction

This chapter looks at the serial data protocols used by the major world automotive manufacturers. It is by no means complete, but it represents some of what has been published in various articles and trade journals during the last several years.

First, Table 17.1 compares and contrasts, at a high level, some generalities about the major standardized protocols. A table like this is in response to those individuals who desire to know "which protocol should I use?" In reality, this is a difficult question. In practice, it probably has already been decided for you. If you are an electronics module supplier, you have to conform to what is already there (unless you are the first node on some new bus). If you are a vehicle manufacturer, the job is simple—invent something new. The "not invented here" (NIH) syndrome is alive and well at the car companies.

The Japanese Society of Automotive Engineers (JSAE) undertook a large project in 1989 through 1991. This effort was to analyze and compare several popular multiplex protocols by scientific methods. These references contain a wealth of information and are recommended reading. Reference [1] gives the test setup and test procedure, and references [2, 3] present the results. The protocols tested were J1850 one-wire, J1850 two-wire, CCD, PALMNET, and CAN.

Table 17.1 is not generally useful because different protocols are for different things. For example, comparing a UART-based protocol with J1939 is not valid because UARTs are mostly for Class A applications, and J1939 is for Class C.

TABLE 17.1
WORLD MULTIPLEX STANDARDIZATION CHART

Protocol	Uses	Advantages	Drawbacks
Class 2	• Class A functions • Class B functions (OBD-II compatible)	• Single wire • VPW encoding • one-chip solutions	• Hard to understand
SCP	• Class A functions • Class B functions (OBD-II compatible)	• Faster speed	• Two wire • Expensive hardware • PWM, lots of edges
ISO 9141	• Class A functions (OBD-II compatible)	• Widely used	• Not intended for multinode systems
CCD	• Class A functions • Class B functions	• Simple	• Proprietary • Limited hardware
J1922+J1708	• Class A functions • Heavy-duty diagnostics • General comm. • Some control	• Simple • Uses SCI port	• Limited capability • Will be replaced by J1939
J1587+J1708	• Class A functions • Heavy duty diagnostics • General comm.	• Simple • Uses SCI port	• Limited scope
ABUS	• Class A functions • Class B functions	• Simple	• No standard approval
VAN	• Class A functions • Class B functions • Class C functions	• Good error capability • Good bit encoding • No license needed • Standard XCVR spec • In-frame response	• Complicated encode • 250 kb/s maximum • Separate XCVR • No wide buy-in
CAN	• Class A functions • Class B functions • Class C functions	• Powerful error capability • NRZ with stuffing • Low noise • Many manufacturers • 1 Mb/s maximum	• No standard XCVR • Separate XCVR • Need license to make
J1939	• Class A functions • Class B functions • Class C functions	• Complete • Good truck standard • Getting wide buy-in	• Unique message strategy

Anyway, the following is a look at the protocols used by the major automakers. If known, the platform(s) and vehicles using the protocol are listed. The protocols used are categorized by class.

A question mark "?" does not indicate that there is not usage for that class; it simply means that there has been no wide public disclosure in that area.

Audi

Class A:

- ?

Class B:

- ABUS
- CAN
- ISO 9141-2 (for diagnostics)

Class C:

- ?

BMW

Class A:

- I-Bus (Instrumentation bus)
    - Connects navigation, audio, keyboard, displays, diagnostics, and so forth
    - Arbitrated
    - Single wire
    - 9600 b/s
    - NRZ
    - Frame = start bit, eight data bits, stop bit, parity bit

- Body electronics bus (P-Bus?)
  - Connects doors, sunroof, relays
  - Star topology
  - Mostly software-driven, dedicated, link
  - Door bus (M-Bus?)
  - Connects main control, and power window switching block
  - Unidirectional
  - Special ASIC
- Local interconnect network (LIN)

Class B:

- Time-Triggered Protocol/B (TTP/B)
- ISO 9141-2 (for diagnostics)
- CAN
  - 530i, 540i, 730i, 740i, 840i, 850i (IP/PT MY93)

Class C:

- Time-Triggered Protocol/C (TTP/C)
- CAN
  - Engine, transmission, throttle control, MY94
  - 16 nodes on high-end cars, MY97
- MOST
- FlexRay

DaimlerChrysler

Class A:

- J2057 and J2058 Information Reports
- LIN

Class B:

- DaimlerChrysler collision detection (CCD, or C^2D) in MY1995 models (average two/vehicle):

TABLE 17.2
CHRYSLER CCD USAGE

Model	ECM	TRANS	BODY	CLUSTR	HVAC	INFO	CONSOL	AIRBAG	ABS	TRIP	THEFT	Total
LHS	X	X	X	X	X	X		X	D			8
N Yorker	X	X	X	X	X	X		X	D			8
Concorde	X	X	X	X	X	X		X	D			8
Intrepid	X	X	X	X	X	X		X	D			8
Vision	X	X	X	X	X	X		X	D			8
Voyager	X	X	X	X		X		X		X		7
Caravan	X	X	X	X		X		X		X		7
Twn&Cn	X	X	X	X		X		X		X		7
Cirrus	X	X	X	X				X				5
Stratus	X	X	X	X				X				5
LeBaron	X	X	X	X				D		X	X	7
Neon								D				1
G Cherokee	X				X	X	X		D		X	5
Cherokee	P										P	2

X = ECU w/CCD chips.
D = CCD chips for diagnostics only.
P = point-to-point.

- VPW J1850 "Carlink"
  - MY94 LH cars, diagnostics only
  - MY97 and later from Harris Semiconductor
  - Complete switch to J1850 (from CCD) in MY02, averaging three chips/vehicle
- ISO 9141-2 for most diagnostic purposes in the future
- Neon 1995

Class C:

- Has investigated and experimented with CAN
- MOST

Citroën

Class A:

- ?

Class B:

- VAN
  - Mil-XM (body, IP MY94)
  - Xantia (MY98)

Class C:

- ?

Daihatsu

Class A:

- Multiplexed body wiring system
  - Connects lamps, wipers, locks, defogger, horn, and buzzers
  - Arbitrated
  - Two wires—transmit and receive
  - 5.21 kb/s
  - NRZ encoding
  - Frame = start bit, 4-bit ID, 3-bit control, parity, stop bit (or data byte, no ID)
  - Master–slave
  - DSC1000, DSC1100, DSC1200 chip set

Class B:

- ?

Class C:

- ?

Fiat

Class A:

- LIN

Class B:

- VAN
  - Lancia, body multiplexing (MY98)

Class C:

- MOST
- CAN

Ford

Class A:

- Audio Control Protocol (ACP) [4, 5]
  - UART-based (J1708 physical layer)
  - J2178 message structure

Class B:

- ISO 9141-2
  - Windstar diagnostics, Mondeo (powertrain [P/T] MY93)
- SCP
  - Mustang, Thunderbird (MY94)
  - Lincoln powertrain, climate control, lighting, driver door/seat (MY95)
  - All powertrain (MY96)
- MSCAN
  - Various (MY02+)

Class C:

- HSCAN
  - Various (MY02+)

General Motors

Class A:

- Class 2
- UART
- E&C

Class B:

- Class 2 (also for diagnostics)
- UART (also for diagnostics)
- GMLAN (CAN) mid-speed and GMLAN low-speed (single-wire CAN [SWC])
  - Various (MY02+)

Class C:

- GMLAN high-speed
- Various (MY02)

Honda

Class A:

- LIN

Class B:

- J1850 (Okidata, PWM)
- Single-wire CAN

Class C:

- MOST

Jaguar

Class A:

- LIN

Class B:

- J1850 (SCP)
- CAN
- Saloon (body, MY96)

Class C:

- CAN

Mazda

Class A:

- ?

Class B:

- Protocol for automotive local area multiplexing network (PALMNET) [6]
  - Supplier is Furakawa (ALPS)
  - Protocol for automotive low- (and medium-) speed network
  - Uses basically the J1850 protocol and custom hardware
  - Cosmos and Eunos models, body, transmission, engine, MY90
  - 929, climate control, MY96

Class C:

- Advanced PALMNET [7]
  - 16-bit CRC, NRZ, up to 1 Mb/s

Mercedes

Class A:

- ?

Class B:

- ISO 9141-2 (for diagnostics)

Class C:

- CAN
    - 1991 S-Class model **FIRST AUTO CAN USAGE**
    - 1992 C-, S-Class models, 1996 E-, SL-Class models
    - Engine/transmission/throttle/traction control
    - Climate control, audio
    - Trucks (dash, engine, transmission, brakes, retarder, chassis)
    - CAN in all car models by MY94
    - 15 nodes 1998 CLK
- D^2B optical bus

Mitsubishi

Class A:

- Mitsubishi Intelligent Cockpit System (MICS), MY90—Diamante
    - Connects seats, mirrors, steering wheel
    - Single-wire, 12-V drive
    - Bus topology
    - 6.7 kb/s
    - Frame = start bit, 12-bit data, 4-bit CRC, stop bit
    - PWM encoding
    - Arbitrated

Class B:

- Proprietary 7.8 kb/s, NRZ, token-passing, one-wire, Diamante MY91

Class C:

- ?

Nissan

Class A:

- IVMS (Hitachi)
  - Sunny, Infiniti, Micra, Stanza, windows/doors/seats, MY91
  - Master–slave, 27.8 kb/s, PWM, single wire
- Body multiplexing bus (with assistance from Hitachi), 1997
  - Connects body controller and as many as 25 local control units (LCUs)
  - Arbitrated
  - Two-wire differential
  - 28 kb/s
  - PWM encoding
  - Frame = SOM (start of message), 2-bit COM (command), 6-bit address, 1 or 2 bytes data, EOM (end of message)
  - Bus topology

Class B:

- Diagnostic bus
  - Connects engine controller, transmission controller, and so forth with diagnostic port
  - Arbitrated
  - Three wires—clock, transmit, and receive (9-volt drive)
  - 9600 b/s
  - NRZ encoding
  - Frame = start bit, 1-byte data, stop bit
  - Bus topology

Class C:

- ?
  - Cedric, Gloria, engine/body/sensors, MY97
- MOST
- CAN

Opel

Class A:

- ?

Class B:

- CAN
- VAN
- J1850
- ISO 9141-2 (for diagnostics)
- Keyword 2000
  - 1996 Vectra (Siemens ECM, Bosch, Delco)
  - 1997 Omega/Catera (Bosch ECM and transmission control module [TCM])

Class C:

- ?

Peugeot

Class A:

- VAN

Class B:

- VAN, body MY96
- CAN

Class C:

- CAN

Renault

Class A:

- VAN
- LIN

Class B:

- VAN
  - Laguna 1994, integrated information display
  - Models 19, Safrane 1995, instrumentation, door, engine, airbags, transmission, ABS
- CAN

Class C:

- CAN
- IDB-1394

Rover

Class A:

- ?

Class B:

- Proprietary
- ISO 9141-2 (for diagnostics since MY90)

Class C:

- CAN, wheel speed systems, sensor sharing devices, MY95

Saab

Class A:

- ?

Class B:

- Proprietary (model 900, body, IP, engine, MY93)
  - "Trionic" engine management system (32-bit CPU)
  - "Sensonic" automatic clutch system
- SAE J1850
- ISO 9141-2 (for diagnostics)

Class C:

- CAN

Toyota

Class A:

- Custom protocol
  - Special assistance by Nippondenso
  - 5 kb/s
  - NRZ-5 encoding
  - Arbitrated
  - Frame = SOF, 2-byte ID/data, 8-bit CRC, end of date (EOD), response, end of frame (EOF), IFR
  - BUS topology
  - Single wire
- LIN

Class B:

- Proprietary
  - One example—a 1983 Century with fiber-optic multiplex link (16 kb/s?)
  - Link-connected door windows, vent windows, door lamps, seats, seat heaters

Class C:

- Special [8, 9]
  - Connects ABS, four-wheel steering (4WS), traction, transmission, throttle, four-wheel drive (4WD), and active suspension
  - 1 Mb/s
  - NRZ-5 encoding
  - Token-passing
  - Frame = SOF, 5-bit address, 3-bit control, 8-bit ID, 0–8 bytes data, 8-bit CRC, EOM, inter-message separation (IMS), idle
  - Bus topology, 20-m maximum length
  - Two-wire differential, shielded media
- "i-Four"
  - Connects ABS, 4WS, traction, transmission, 4WD, and active suspension
- J1850 (41.6 kb/s, PWM)
  - Two-wire differential, shielded media
- Body electronic area network (BEAN) [10, 11]

Volkswagen

Class A:

- LIN

Class B:

- ABUS
- ISO 9141-2 (for diagnostics)
- CAN, Passat MY98, doors, seat, central, 62.5 kb/s, 5–6 nodes
  - Expanding to Golf, Skoda Octavia

Class C:

- CAN
- MOST
- TTP

Volvo

Class A:

- ?

Class B:

- ISO 9141-2, Model 850 turbo, engine, MY94
- VAN, Model 850, IP, MY96

Class C:

- CAN, steering, MY99

Other Miscellaneous Protocols/Hardware

A few other protocols are worth mentioning, either because some vehicle manufacturer is producing them or they were proposed as an industry standard at one time.

Wiper Control System [12]:

- United Technologies Automotive

Distributed real-time control system [13]:

- Xerox Corp.

Multiplexed data link with token passing [14]:

- Allied Signal Corp.

Salplex Class A multiplex system [15]:

- Salplex Ltd.

AUTOLAN [16]:

- General Instrument Corp.
- Class A, B, or C
- Up to 4 Mb/s
- Autolan controller IC

Multiplex protocol for medium-scale systems [17]:

- Yazaki Parts Co. Ltd.
- Class B
- 62.5 kb/s
- Arbitrated
- Two-wire differential
- PWM encoding
- Frame = start bit, 1-byte data, stop bit
- Bus topology
- Tested on 1987 Japanese luxury sedan

Standard Army Vetronics Architecture (SAVA) [18]:

- U.S. Army Vetronics Technology Center
- Land vehicle multiplexing

Time-Triggered Protocol (TTP) [19–21, 23–25]:

- Technical University of Vienna
- Real-time, universal time clock

IP Bus [22]:

- Interactive Pioneer Bus (in conjunction with NEC)
- For entertainment unit multiplexing
- Various bit rates to 26 kb/s and up to 128 bytes/message over STP
- PWM, arbitrated, two-wire
- In use since MY92

Optical LAN:

- Communication and Control (C&C) Electronics, Ltd., U.K.
- For Yazaki; D2B protocol.

References

[1] Hiwa, N., and Nobutoki, Y. "JSAE Recommended Practice—Evaluating Method of the Automotive Distributed Control Multiplex Systems." SAE Paper No. 910716, Society of Automotive Engineers, Warrendale, PA, 1991.

[2] Hiwa, N., and Fujiki, N. "An Evaluation of Multiplexing System for Automotive Distributed Control." SAE Paper No. 910718, Society of Automotive Engineers, Warrendale, PA, 1991.

[3] Fujiki, N., and Hiwa, N. "Final Report on Evaluation of Multiplexing Systems for Automotive Distributed Control." SAE Paper No. 920228, Society of Automotive Engineers, Warrendale, PA, 1992.

[4] Emaus, B. "ACP—Ford's Audio Control Protocol." SAE Paper No. 940142, Society of Automotive Engineers, Warrendale, PA, 1994.

[5] Emaus, B., and Worcester, D. "An Evaluation of ACP Collision Resolution Performance." SAE Paper No. 952578, Society of Automotive Engineers, Warrendale, PA, 1995.

[6] Inoue, T., and Himono, Y. "Protocol for Automotive Local Area Network (PALMNET)—A Newly Developed In-Vehicle Communication System Based on SAE J1850." SAE Paper No. 890535, Society of Automotive Engineers, Warrendale, PA, 1989.

[7] Kimura, H., and Hiwa, N. "The Development of the Advanced Protocol for Automotive Local Area Multiplexing Network (Advanced PALMNET)." SAE Paper No. 940365, Society of Automotive Engineers, Warrendale, PA, 1994.

[8] Nakamura, S., Isobe, T., and Hirabayashi, Y. "The High Speed In-Vehicle Network of Integrated Control System for Vehicle Dynamics." SAE Paper No. 910463, Society of Automotive Engineers, Warrendale, PA, 1991.

[9] Azuma, M., Takagi, Y., Chujo, N., and Kawahashi, A. "Development of a Class C Multiplex Control IC." SAE Paper No. 930003, Society of Automotive Engineers, Warrendale, PA, 1993

[10] Isobe, T., Honda, H., Uehara, S., and Akiyama, S. "A Low-Speed In-Vehicle Network for Body Electronics." SAE Paper No. 920231, Society of Automotive Engineers, Warrendale, PA, 1992.

[11] Shinojima, Y., Kawakami, M., and Kawahahi, A. "A Multiplexing Communication IC for Automotive Body-Electronic Control." SAE Paper No. 940364, Society of Automotive Engineers, Warrendale, PA, 1994.

[12] Juzswik, D., and Chin, K. "Microcomputer Based Serial Multiplex Wiper Control System." SAE Paper No. 860393, Society of Automotive Engineers, Warrendale, PA, 1986.

[13] Bonikowski, G., Gottwald, E., and Liu, S. "Xerox's Distributed Real Time Control System." SAE Paper No. 870517, Society of Automotive Engineers, Warrendale, PA, 1987.

[14] Pattantyus, T. "Multiplexed Data Link with Token Passing." SAE Paper No. 870518, Society of Automotive Engineers, Warrendale, PA, 1987.

[15] Betts, W. "Salplex Class A Multiplex System—A Practical Application." SAE Paper No. 890538, Society of Automotive Engineers, Warrendale, PA, 1989.

[16] Harris, J., and Bailey, N. "General Instrument AUTOLAN—A High Speed Multiplexing System for Automotive Body Wiring Systems." SAE Paper No. 890541, Society of Automotive Engineers, Warrendale, PA, 1989.

[17] Nagasawa, K., and Uda, M. "A Multiplex Protocol for Medium Scale Systems." SAE Paper No. 890543, Society of Automotive Engineers, Warrendale, PA, 1989.

[18] Richardson, P. "Distributed Communications in Embedded Military Systems." SAE Paper No. 940139, Society of Automotive Engineers, Warrendale, PA, 1994.

[19] Kopetz, H. "Fault Management in the Time Triggered Protocol (TTP)." SAE Paper No. 940140, Society of Automotive Engineers, Warrendale, PA, 1994.

[20] Kruger, A., and Kopetz, H. "A Network Controller Interface for a Time-Triggered Protocol." SAE Paper No. 952576, Society of Automotive Engineers, Warrendale, PA, 1995.

[21] Kopetz, H., Hexel, R., Kruger, A., and Millinger, D. "A Synchronization Strategy for a TTP/C Controller." SAE Paper No. 960120, Society of Automotive Engineers, Warrendale, PA, 1996.

[22] Ikata, Y., and Kurosaki, M. "The Development of the Interactive Pioneer Bus (IP-BUS)." SAE Paper No. 960122, Society of Automotive Engineers, Warrendale, PA, 1996.

[23] Kopetz, H., Hexel, R., and Kruger, A. "A Prototype Implementation of a TTP/C Controller." SAE Paper No. 970296, Society of Automotive Engineers, Warrendale, PA, 1997.

[24] Kopetz, H. "TTP—A New Approach to Solving the Interoperability Problem of Independently Developed ECUs." SAE Paper No. 981107, Society of Automotive Engineers, Warrendale, PA, 1998.

[25] Hedenetz, B., and Belschner, R. "Brake-by-Wire Without Mechanical Backup by Using a TTP-Communication Network." SAE Paper No. 981109, Society of Automotive Engineers, Warrendale, PA, 1998.

Chapter 18

Manufacturing Automation

Introduction

Besides in-vehicle use, serial data networking has become extremely widespread in other areas, especially in manufacturing automation. It is projected that eventually there will be more CAN nodes implemented for this purpose than for automotive purposes.

Obviously, there are many more uses for this technology than manufacturing automation. CAN and other protocols are quietly being instituted in medical equipment, office and home security systems, agriculture, and so forth.

Popular Protocols

The following is an overview of the better-known manufacturing automation serial data protocols, in alphabetical order.

ARCNET

A company called Datapoint in 1977 developed ARCNET, or attached resource computer network. By 1987, the ARCNET Trade Association formed to promote the protocol. ARCNET is now ANSI 878.1 standard.

Some basic attributes of the protocol are that it operates at 2.5, 5, or 7.5 Mb/s. It can transfer 1 to 507 bytes of data via a token-passing scheme.

By May 1996, there were 7 million nodes installed worldwide. There are reportedly a couple of automotive navigation applications by Toyota and BMW. Standard Microsystems Corporation developed an interface IC named the COM20023.

Actuator Sensor Interface

The actuator sensor interface (ASI) is also referred to as "ozzie." It is the result of an eleven-company program (mostly German) for low-end, low-cost applications. A third-party entity, AS-International, administers the protocol now.

ASI runs at 167 kHz, master–slave, with as many as 31 nodes per master and 100 meters maximum bus length.

DeviceNet

DeviceNet [1] is an interesting manufacturing automation protocol because it is based on CAN. It is not the only one, however. SDS (see later) is also CAN-derived. DeviceNet has about 47% of the U.S. market and was developed by Allen Bradley in 1994. The Open DeviceNet Vendor Association (ODVA) handles the protocol now. ODVA releases specification updates, schedules trade shows, forms working committees, and so forth.

Hitachi, Omron, and Cutler-Hammer are members of ODVA. Beginning with about 100,000 installed nodes in 1994, DeviceNet had about 3.7 million nodes in use in 1999. General Motors announced that its North American Operations (NAO) adopted DeviceNet as its in-plant automation standard [2].

DeviceNet can run at 125 kb/s over 500 meters, at 500 kb/s with 128 nodes over 90 meters, or at 1 Mb/s with 64 nodes over 23 meters.

LONWORKS

In the late 1980s, Echelon developed a protocol called LONWORKS. A custom chip from Motorola and Toshiba called the "Neuron IC" performs the bus interface function. LONWORKS works at 1.25 Mb/s with a large number of potential nodes (thousands). There are about 1500 users and more than one million nodes installed at this time. Applications include building security and utility-monitoring systems.

LONWORKS is well suited for use as a "sensorbus" in manufacturing automation (see later in this chapter). In this case, the sensorbus would interface low-end nodes to a higher-end backbone such as Ethernet, TCP/IP, I^2C, or Seriplex.

Profibus

Profibus (process field bus) is mostly a German–U.S. collaboration. General Motors is one of about 130 subscribers. GM has indicated that it intends to use Profibus for U.S. powertrain and general European industrial applications.

ICs are available from Motorola, Intel, Siemens, Philips, ShipStar, and SMAR. One of the most recent ASIC from Siemens runs at the new Profibus speed of 12 Mb/s instead of the usual 500 kb/s.

Profibus has about 16% of the U.S. market for industrial automation serial interfaces.

SDS

SDS is another CAN-based industrial protocol [3, 4]. It was developed by Honeywell's Micro Switch Division and implements object-oriented software. SDS has about 26% of the U.S. market with about 100,000 nodes installed. A General Motors engine plant in Australia and a Crown Cork & Seal aerosol can plant in the United States use SDS.

Other Industrial Networks

Some other minor players in the industrial automation and general commercial sector are briefly mentioned here.

Bitbus

- Intel 1984; 8044 microprocessor; distributed OS; multitasking; RS-485

Fieldbus Foundation

- Up to 2.5 Mb/s; token-slot, 240 nodes/segment

GPIB (General Purpose Interface Bus)

- IEEE-488, 1978; data acquisition

HART (Highway Addressable Remote Transducer)

- Rosemount; two-way communication atop traditional 4–20 mA loops

Interbus S

- Phoenix Contact, sensor/actuator ring bus; RS-422; Europe
- 500 kb/s; 40 m; ring with master

Seriplex

- Automation Process Control Co.; ASIC-based; bit-level
- Bought by Square D
- 500 kb/s; 256 nodes

Categorizing the Industrial Buses

Industrial automation serial data protocols can be categorized into three main classes. The slower, cheaper ones that do simple logic control over short distances are called "sensorbuses." The fast, expensive protocols suited for process control over long distances at great speeds are called "fieldbuses" [5]. In between are "devicebuses." A graphical depiction of this is shown in Fig. 18.1.

A sensorbus has nodes such as switches and valves and communicates by a simple bit-level strategy. A devicebus may contain programmable logic control (PLC) or PCs, and so forth, and communicates via a byte format. Fieldbuses

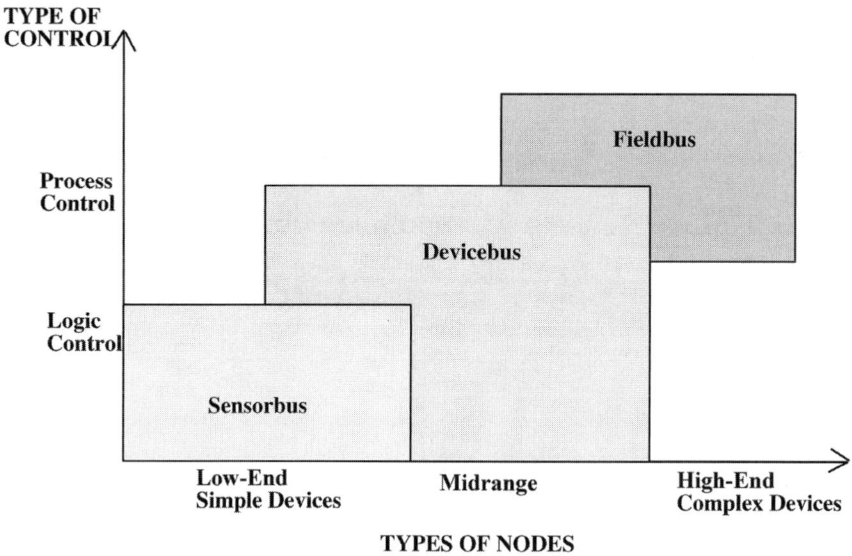

Fig. 18.1 Industrial bus classification.

usually connect the plant's main control systems together and have a complex message-level communication format.

Protocols that could be classified as sensorbus members include the following:

- DeviceNet
- Seriplex
- ASI
- LONWORKS

These are devicebus protocols:

- DeviceNet
- LONWORKS
- SDS
- Interbus S
- Profibus-DP

These are examples of fieldbuses:

- Profibus-PA
- WorldFIP
- Fieldbus Foundation

Note that there is overlap in this classification scheme. For example, DeviceNet is considered both a sensorbus and devicebus protocol.

References

[1] DeviceNet Application Protocol. Noonen, Siegel (CiA).

[2] The Hansen Report. March 1997

[3] SDS: A CAN Protocol for Plant Floor Control. Crovella (CiA).

[4] SDS: A CAN Protocol for Plant Floor Control II. Beck (CiA).

[5] FieldBus—The Bottom-Up Approach to an Open DCS. Phinney, Honeywell.

Chapter 19

Industry Activities

Introduction

This chapter discusses certain activities coordinated by SAE and other groups to develop and promote multiplexing in vehicles.

USCAR

The United States Council for Automotive Research (USCAR) was formed on June 8, 1992 to study and develop advanced vehicle technologies. At this time, it is composed of at least the following consortia or partnerships:

- U.S. Advanced Battery Consortium includes the U.S. Department of Energy and the Electric Power Research Institute
- Computer aided design/computer aided manufacturing (CAD/CAM) group
- Low Emissions Paint Consortium
- Environmental Research Consortium includes EPA and Michigan state officials
- Auto/Oil Air Quality Improvement Research Consortium includes oil company members
- Electric Vehicle Study Group
- Supercomputer Automotive Applications Partnership
- Partnership for a New Generation of Vehicles (PNGV or "supercar")

- Other groups:

  Automotive composites
  Vehicle recycling
  Occupant safety research
  Electrical Wiring Component Application Partnership (EWCAP)

USCAR is supported mainly by SAE and the American Automobile Manufacturers Association (AAMA).

As a historical note, it is interesting to mention the activities of a now extinct USCAR partnership that was involved in Class C multiplexing, the High Speed Serial Data Communications Partnership Protocol Work Group (HSSDCPPWG). Even the acronym needs an acronym, so the group usually went by the name "Data Links Partnership" or DLP. This group was formed in May 1991 before the advent of USCAR.

The DLP was motivated by the impending need for a high-speed serial data communication standard for vehicles in the 1990s and beyond. There were many unanswered questions concerning Class C protocols and applications. These questions needed to be answered and a standard protocol made available before the applications arrived. The major driving force behind the DLP was the desire not to repeat the mistakes of Class B protocol standardization. For example, many people categorized J1850 as a disaster, mainly because of its "multiple standards within a standard" realization. Members of the DLP hoped that such a debacle could be avoided with Class C and that an early consensus could be gained on a protocol standard among the "Big Three" automakers.

The objectives of the DLP included identifying the important characteristics that define an ideal Class C protocol and characterizing available protocols in terms of the ideal. The group made a lot of progress in the years 1991 through 1993. The DLP met monthly. Sometimes, there were 20 attendees or more (there were about 14 official members).

Quality function deployment (QFD) was the method used in determining the ideal protocol attributes. QFD is based on customer needs and system characteristics that will achieve those customer needs. The results of the first part of the DLP's objectives was a list of ideal Class C system attribute categories:

- Response time
- Network throughput/efficiency
- Probability of undetected errors
- Addressing strategy
- Number of single-point failures
- Error detection
- Error response
- Media access
- Types of frame/header elements

Ultimately, three reports were written and released by the group. Unfortunately, these reports are confidential and are for only GM, Ford, and DaimlerChrysler. They could be made available to any company by one of the three authoring entities under a nondisclosure agreement. The names and dates of the documents are as follows:

- Determination of the Ideal Protocol, January 23, 1993
- Comparison of Candidate Protocols to the Ideal Targets, July 16, 1003
- Evaluation of Transmission Media Alternatives, November 18, 1993
- Shielded twisted pair and plastic fiber optic only

The DLP quietly disbanded in December 1993. At the end, there was dissension and disagreement about the direction that future work should take. Several disparaging comments were made by group members, such as "We can live without high speed multiplexing networks for now," and "Nothing absolutely requires a bit rate greater than 41.6 kb/s." Another was that "One can design around the need for multiplex in large data sharing applications." This can be done by integrating functions into fewer boxes and by integrating data into fewer messages (data packing). Someone stated that future applications such as intelligent vehicle highway systems (IVHS) and the "mobile office" will drive a need for Class C links.

One of the official conclusions of the DLP was that CAN 2.0 B would probably fit the needs of Class C applications in MY00 and beyond. The media used would be shielded twisted pair or fiber optics. In hindsight, one can see that the media prediction was wrong. The mass use of fiber optics for data communications is still several years away. Unshielded twisted pair is the medium of choice for most automotive Class B and Class C bus applications.

SAE High-Speed CAN Implementation Task Force

A little more than a year after the demise of the USCAR DLP, it was decided that the time was right to develop a U.S. passenger car CAN standard. A new SAE subcommittee of the multiplex group was formed and subsequently named the SAE High-Speed CAN Implementation Task Force. This task force took the number J2284 and had its first meeting on April 20th 1995.

The high-speed task force had many working members, including vehicle OEMs as well as their main suppliers such as Delphi Delco Electronics Systems. Development tool and scan tool manufacturers and IC vendors were active.

An early decision was made to coordinate the rollout of Class C multiplex applications and the bus they would use. CAN would be the protocol of choice, and some of the multiple-protocol mistakes of J1850 would be avoided. The goal was a high-speed standard for passenger cars, so a bit rate of 500 kb/s was chosen. The medium chosen was unshielded twisted pair over a maximum bus length of 30 meters and 16 nodes.

The J2284 document also included some connector requirements as well as bit construction guidelines for maintaining proper bit timing/synchronization.

It has been asked, "Why develop a new standard? . . . Why not adopt ISO 11898 or J1939?!" ISO 11898 was referenced in J2284, although not used as is, primarily because ISO 11898 does not specify a bit rate. J2284 is actually intended to be ISO 11898 but with supporting requirements for 500 kb/s. J1939 was not used because it is too specific for truck and bus applications. J2284 was to be only for passenger vehicles. The J2284 group also decided that 29-bit message IDs were not necessary. No good argument could be put forth to require the extra overhead and throughput to support the extended IDs. On a 500-kb/s bus, most data would be packed into a small number of periodic messages; hence, the extra messages the 29-bit ID provides are not useful.

Wiring Partnership

One of the latest USCAR groups is the Electrical Wiring Component Application Partnership (EWCAP). Its purpose is to look for ways to reduce the number of electrical design variations in automotive electrical systems. These are some of its goals:

- Shrink the number of variations from 100+ to <10 within 20 years
- Reduce domestic industry's inventory of connector part numbers by 80% from 15,000+ to 3000
- Reduce connector families from 47 to 8
- Make diagnosing and repairs easier
- Increase system reliability

An interesting comment published by the group was that

". . . automakers view multiplexing . . . as only a limited answer to the problem of wire proliferation."

Six full-time engineers support EWCAP. These engineers are volunteers, two each from GM, Ford, and DaimlerChrysler. There is an annual budget of $750,000.

A first specification, USCAR 1, was written and agreed to by the U.S. Big Three automakers as of April 1996. Implementation was to take place beginning with the 1998 model year and by MY09, was to encompass 90 to 95% of all connectors in use. Four families of terminals are outlined: 1 mm, 1.5 mm, 2.8 mm, and 1/4 in.

CAN in Automation

The industry group CAN in Automation (CiA) was formed in March 1992 as an international users and manufacturers organization. It has a wide variety of members, including the following:

- Philips, National Semiconductor, Motorola, and so forth
- Rockwell Automation
- I+ME, Dearborn Group
- Daimler-Benz AG

There were 370 members in January 1, 2001. CiA is made up of Technical and Business committees. Within these committees are Working Groups and Special Interest Groups endorsed by the CiA General Assembly.

One goal of CiA is to provide technical, product, and marketing information for CAN. Another is to enhance, promote, and advance CAN generally. CiA also wants to develop standards for the physical layer, mostly higher layers, and applications of CAN. Applications include trucks, any "machines on wheels," industrial applications, and medical applications.

CiA has been sponsoring an annual technical event called "The International CAN Conference." The fourth one was held in Las Vegas in 2002. These conferences are always well-attended and contain many CAN technical papers, demonstrations, and product shows. All applications are represented (especially automotive and industrial automation), and there is a good showing of IC vendors and other support entities.

The CiA sponsors an excellent trade journal, to which one may subscribe. For more information, contact PZ Marketing at +49-9131-601091 phone or +49-911-3067283 fax. Or visit the CiA web site at www.can-cia.de.

OSEK/VDX

A large (mostly European) effort is underway to standardize vehicle electronic architectures. Sometimes this work is misinterpreted as a software operating system. In reality, it is a new concept for automotive electronics.

OSEK/VDX is an acronym for "Offene Systeme und deren Schnittstellen fur die Elektronik im Kraftfahrzeug/Vehicle Distributed Executive" [1]. It is more commonly referred to as "Open Systems and Interfaces for Distributed Electronics in Cars." The University of Karlsruhe in Germany coordinates the development of specifications. Member companies in the effort include the following (since 1993):

- Opel
- BMW
- Daimler-Benz
- Mercedes-Benz
- Bosch
- Siemens
- Volkswagen
- PSA–Renault

It is estimated that this specification requires 7.5 kilobytes of ROM and 500 bytes of RAM for implementation.

The main goal of OSEK/VDX is to specify an open architecture for vehicle control and communications that results in a standard for cost-effective integration and portability. The aims are to specify uniform services, interfaces, and protocols for the communication, network management, and operating system of a distributed real-time application. The tie to our discussion is that CAN has been promoted as a solution for the data link layer.

The OSEK architecture is comprised of three layers: communication (data exchange within and between control units), network management (configuration determination and monitoring), and real-time executive (operating system for control unit software).

ITS Data Bus

The Intelligent Transportation Systems (ITS) Data Bus (IDB) Committee was formed in 1995 after a need was identified to develop an open, industry standard in-vehicle bus that would allow installing ITS and other consumer electronics devices in a vehicle, at any time during its life cycle, and would require little or no expert assistance. The IDB is part of the SAE ITS Division. Members include representatives from GM, Ford, DaimlerChrysler, Motorola, Sony, Nokia, and so forth. Cooperation is being provided by the Cellular Telecommunications Industry Association (CTIA).

IDB is supposed to facilitate a "plug-and-play," that is, hot-pluggable, network within vehicles. This network would connect a phone, navigation system, pager, traffic information, personal computer, Internet access, and so forth. Devices will be self-configuring and communicate at an initial data rate of 250 kb/s using a two-wire CAN physical layer (IDB-C). The IDB will make it easier to upgrade vehicle electronics. Because of long vehicle development cycles, the onboard electronics tend to be three to four years behind current technology. Devices on the IDB can be replaced as newer versions become available. Vehicle driver customization and personalization will be possible via the IDB.

A secure gateway will be provided to the rest of the vehicle's electronics. The committee identified the following four standards that needed to be developed:

- J2355 ITS Data Bus Architecture Reference Model and Functional Requirements

  This Information Report describes the overall structure, functions, interfaces, and protocols for the IDB. It also captures the functional requirements that have and will continue to drive the technical development of the standards. These functional requirements were developed at two workshops; the first was held at DaimlerChrysler's Technical Center in Auburn Hills, Michigan, in the winter of 1995, and the second at Honda's R&D facility in Torrance, California, in the summer of 1996. This document has been successfully balloted by the IDB Committee and was published in 1997.

- J2366 IDB Protocol

  This standard will be in three parts, each developed by a subcommittee.

The Physical Layer Subcommittee (J2366/1-IDB Physical Layer) has resolved many of the technical requirements and is working on its first draft of the standard. Some open issues related to EMI/RFI and connector selection are expected to be addressed in the coming months.

The Link Layer Subcommittee (J2366/2-IDB Link Layer Protocol) has developed a draft of the link layer protocol, which is now under subcommittee review. A state diagram has been developed, and the self-configuration and discovery mechanisms have been largely worked out. The detailed design is progressing.

The Application Layer Subcommittee (J2366/3-IDB Application Layer Protocol) has developed a high-level design document, which is currently under subcommittee review. Eventually, this document will contain the message sets required for devices to interoperate on the IDB. It is anticipated that this document will be revised regularly, as new functions and features are incorporated into the IDB.

- J2367 IDB Gateway

This document describes the operation of the IDB to the vehicle system's gateway and defines the minimal set of messages that must be propagated across the gateway so that the overall system is "IDB compliant." This ensures that the gateway, which is connected directly to the vehicle's multiplex bus or other subsystems, does not interfere in any way with the safe and normal operation of the vehicle.

- J2368 IDB Conformance Test

As previously stated, it is important to ensure that the IDB gateway and IDB connected devices do not interfere in any way with the safe operation of the vehicle. This standard will define the test procedure to guarantee that a device is "IDB compliant." This subcommittee is the newest of the IDB task forces.

The IDB committee activities are being partially funded by the Federal Highway Administration. Some of the funds are being used to hire contractors to develop the technical documents, thus enabling the committee to progress more quickly.

Automotive PC Bus

A relatively new idea in aftermarket automotive electronics is that of an "automotive PC bus" [2]. This is the possibility of making one's car an extension of the PC at home. Most, or all, of the features of the home computer could be made available in the person's vehicle(s). Besides the usual features such as Internet access, e-mail, and fax/paging, other features such as navigation, cell phone, weather/traffic information, and security would be integrated into the system. Games and other entertainment could be provided for nondriver occupants.

This bus would be separate from the other electronics in the vehicle. If a gateway is present, it would be a very secure port.

Two proposals are being marketed to car companies today. One of these is called "Connected Car-PC" and is being developed by Intel (Visteon, Citroën, and Volkswagen are partners). Target customers are OEMs and the aftermarket. This system is based on a 32-bit CPU (probably Pentium) and would run Windows® CE. The other proposal is from Microsoft and is basically the same idea as Intel's. At this time, a big drawback to these proposals is price. Windows licenses are not cheap. In fact, the total cost to add one of these networks to a vehicle is about $2000—far more than anyone will pay.

Clarion scheduled "mass production" of the first AutoPC product in the third quarter of 1998. This 1-DIN size unit incorporates Windows CE, a Hitachi SH3 microprocessor, Universal Serial Bus (USB), and 200 voice-recognition commands.

It is distinctly possible that a company like Microsoft may have its sights on acquiring more of the in-car electronics business than only that of the mobile office. This is something that established vehicle electronics suppliers will be watching closely. Pricing of components will be a handicap. As an example, Clarion product costs are approximately as follows:

Clarion AutoPC:

- Base unit: $1299
- CD changer: $375
- Security system: $295
- Global positioning system (GPS) kit: $175

- FM wireless: $49
- Cell phone kit: $195
- Flash card: $199
- Navtec map data: N/A
- Mayday per month: N/A
- Traffic info: $5/month

The total cost for a typical installation would be about $1868 to $2163.

Connected Car-PC has no known production applications at this time, but a show car was presented at the winter 1998 Consumer Electronics Show in Las Vegas. Intel was the main developer of a Ford Expedition outfitted with a Pentium microprocessor with MMX and Windows CE. Some of the features of this vehicle were navigation, Internet access, and DVD movies and games for backseat passengers.

References

[1] Kiencki, U., and Graf, J. "Open Systems and Interfaces for Distributed Electronics in Cars (OSEK)." SAE Paper No. 950291, Society of Automotive Engineers, Warrendale, PA, 1995.

[2] The Hansen Report. March 1997.

Chapter 20

Future Trends

Introduction

Multiplexing will be in development for several years to come and will be used for many years. Some trends are apparent as we approach the maturity of multiplexing. This chapter presents some insight into what will happen in the next few years.

Migration from Standalone to Up-Integrated

In the past, the hardware interface for a multiplexed node involved a standalone IC or ICs. Typically, there was a digital IC and an analog (transceiver) IC. In some cases, these were combined into one chip, such as the MC68HC56, shown in Fig. 20.1.

After about the 1998 model year, most applications are being built with multiplex interfaces that have at least the digital portion up-integrated onboard the microcontroller. Figure 20.2 is an illustration of this where both the digital and analog functions are included on a single-chip design. As in the standalone approach, the transceiver is often a separate IC. Most IC vendors cannot justify complicating the manufacture of the chip by adding analog features. For some serial data link protocols, the transceiver requires battery voltage and thus all of its attendant transient protection circuitry.

Multiple Networks

Most multiplexed vehicles of the future will contain multiple buses or networks [1]. This was covered in Chapter 16.

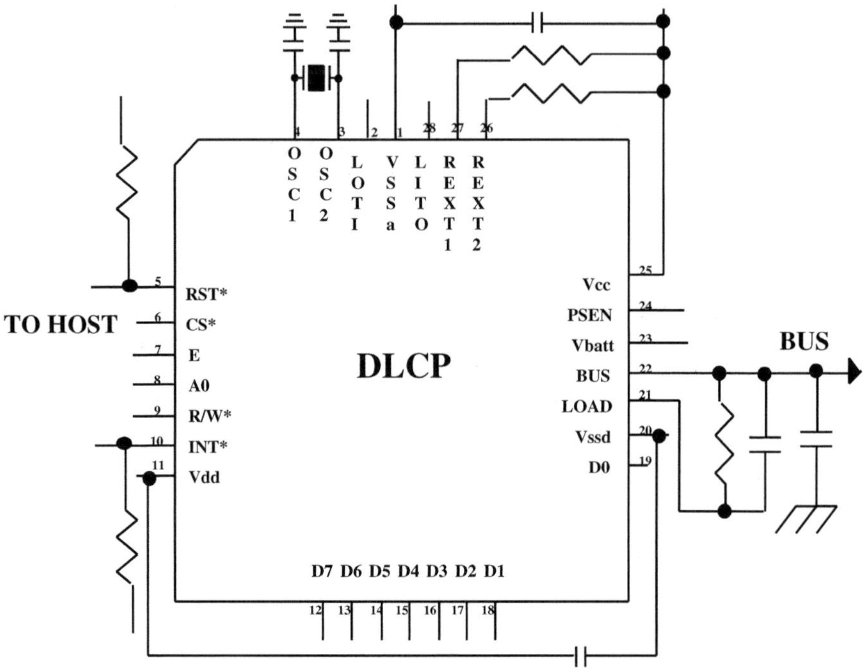

Fig. 20.1 Typical standalone mechanization.

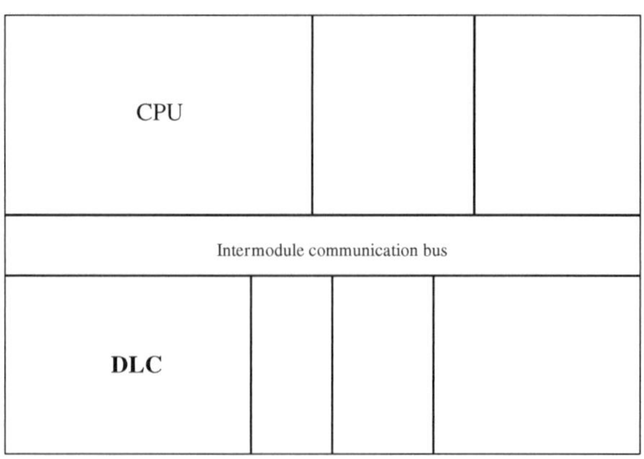

Fig. 20.2 Typical up-integrated mechanization.

Smart Switches/Sensors

The concept of "smart" switches and sensors is not new. This idea of putting electronics at or within a switch, sensor, actuator, and so forth has been a goal of automotive electronic architects for some time. Sometimes, this is referred to as "load control multiplexing" or "smart connector."

As an example, in the conventional approach of connecting switches and loads that are separated by a door panel, separate wires must be run to and from the switches in the door. Power and maybe ground must be run, and as Fig. 20.3 indicates, there is a large wire bundle that must pass through the door hinge area. Especially troublesome are the individual wires from each switch to its load—in this case, probably a seat motor actuator/driver.

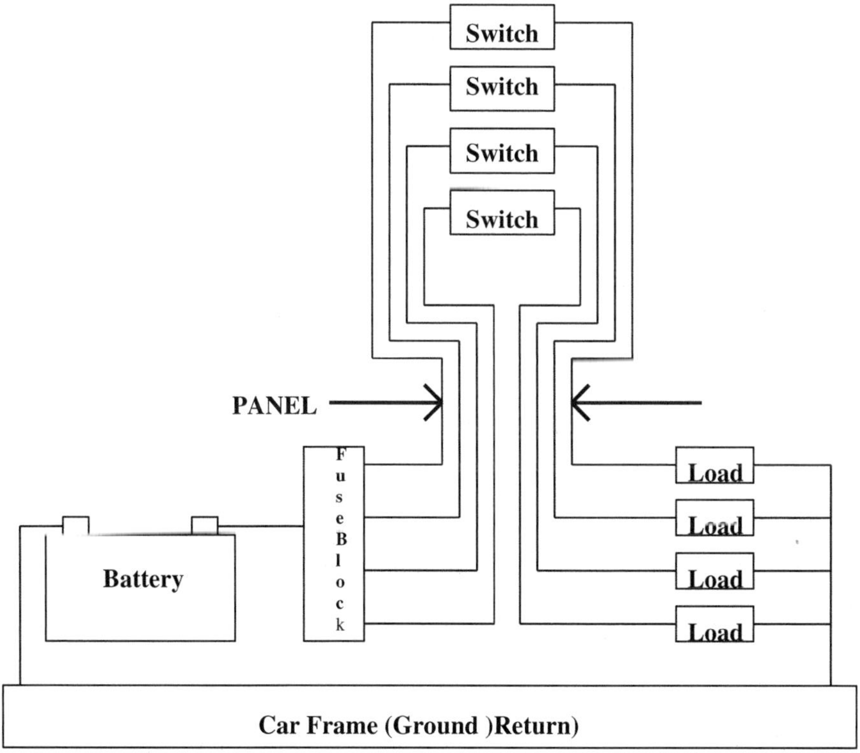

Fig. 20.3 Conventional load control in a door.

Figure 20.4 has another look at conventional load control, showing the large bundle of wires (33 in this case) going into the door.

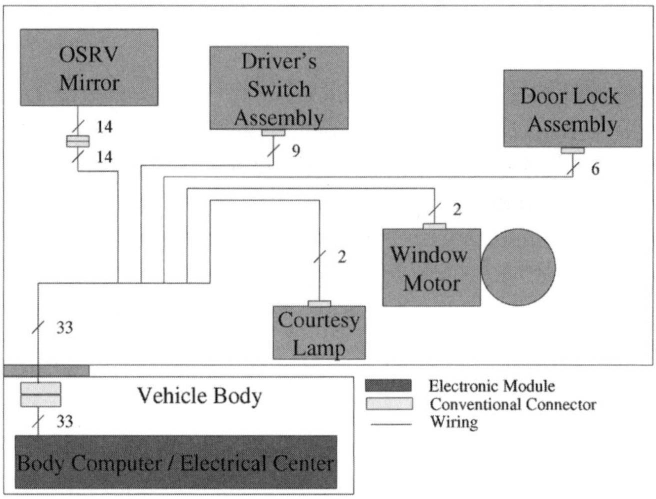

Fig. 20.4 Conventional load control in a door.

In Fig. 20.5, a multiplexed bus has replaced the wires from the switches to the loads. This decreases the size of the wire bundle going through the door hinge. But now there are "smarts" (electronics) added to each switch and load to interface them with the bus. On a pure cost basis, this probably does not make sense. From the viewpoint of weight, size, serviceability, and diagnosability, it makes a lot of sense. But try and convince your purchasing department of this!

Figure 20.6 shows an alternate arrangement, where each individual function in the door does not have an interface to the bus. Rather, a "door module" manages the inputs and outputs and transfers multiplex information to the body computer.

A true "smart connector" architecture would be as shown in Fig. 20.7. In this case, each function in each door is directly connected to the smart subbus.

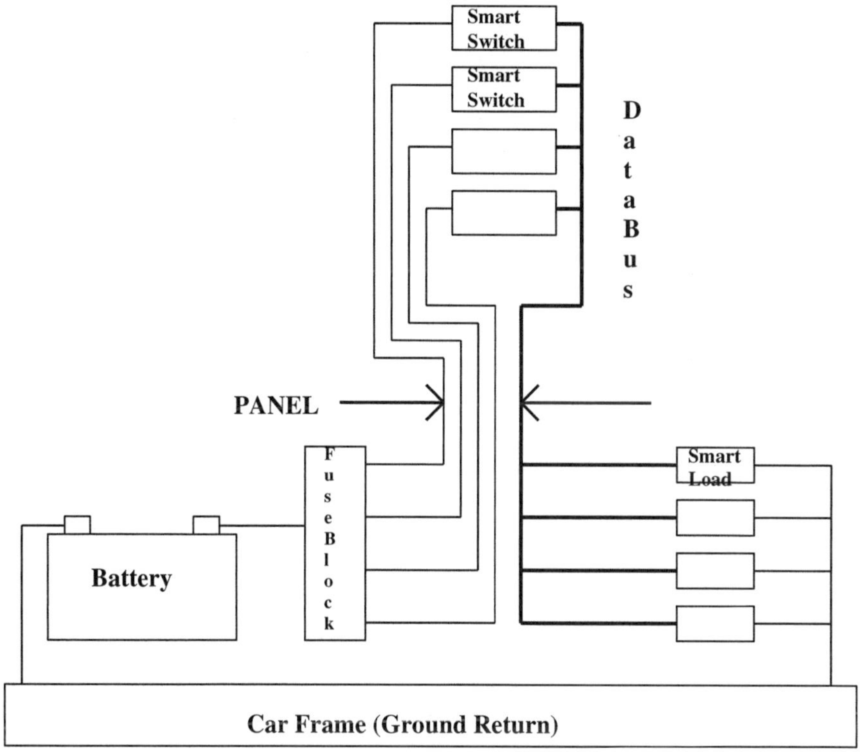

Fig. 20.5 Load control multiplexing in a door.

Load control multiplexing can be applied to the whole vehicle and may be combined with other control functions on the same bus. No doubt, a practical solution will use two or more protocols and distinct networks.

In the conventional approach, each ECU, switch, sensor, and so forth is connected to something else to receive or send data (Fig. 20.8).

In the "all-vehicle load control multiplexing" approach (Fig. 20.9), a common bus connects all data producers and users.

In all-vehicle load control multiplexing, there is a drastic reduction in the number of wires running throughout the vehicle. As in the load control-only

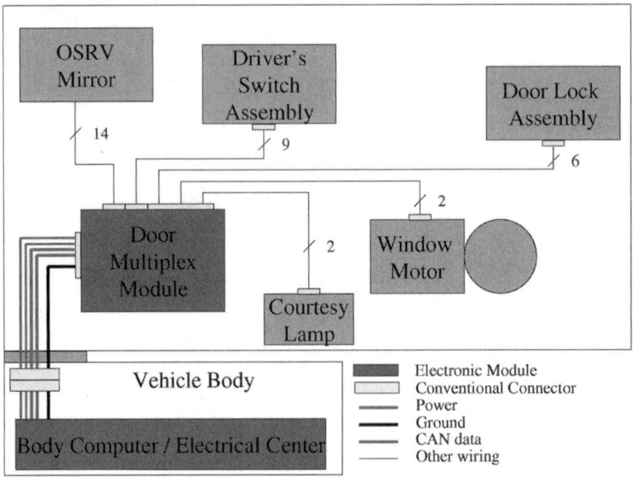

Fig. 20.6 Load control multiplexing in a door, intermediate solution.

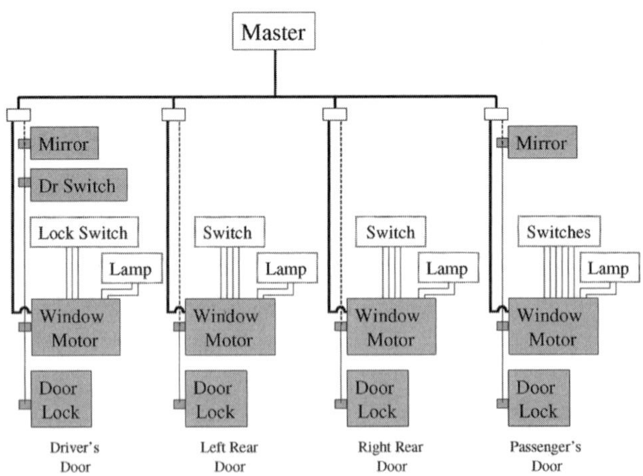

Fig. 20.7 Load control multiplexing in all four doors.

approach, circuitry has to be added to each node to allow interfacing to/from the common bus (shown as a ring in this case).

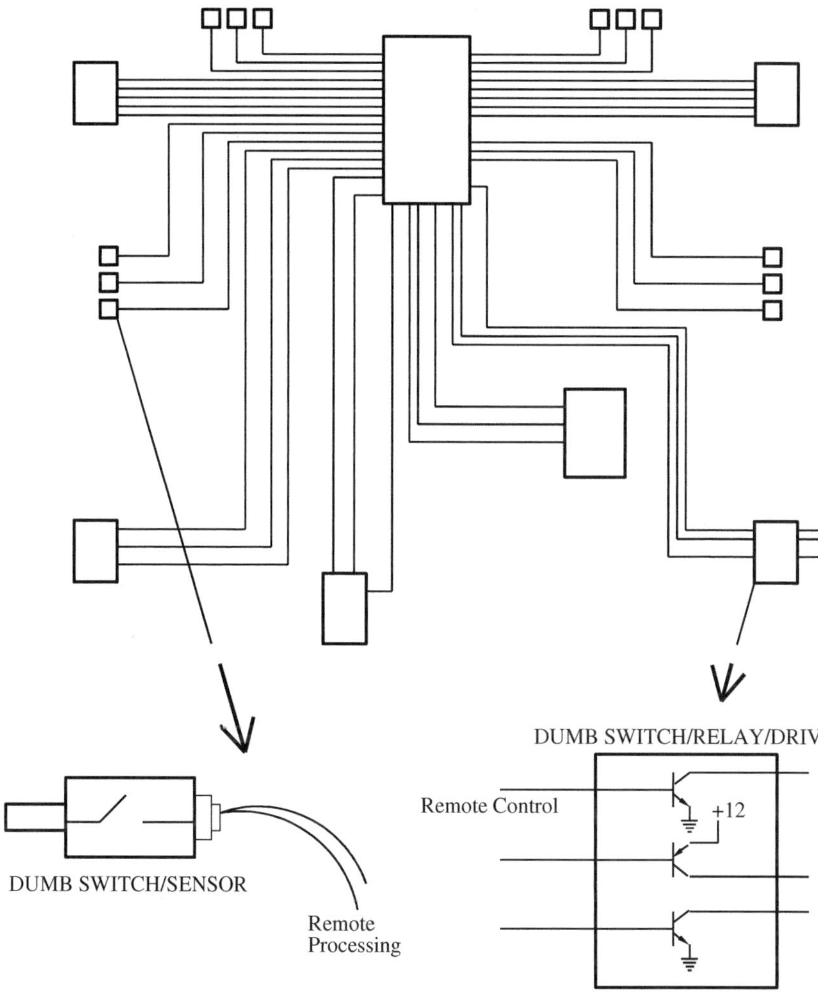

Fig. 20.8 All-vehicle conventional load control.

Generic Hardware

Another design factor that will affect vehicle multiplexing in the future is the concept of generic hardware. Vehicle manufacturers would like to decrease the amount of custom electronics. Using more "off-the-shelf" ECUs, switches, sensors, and so forth would save money, design time, and integration time.

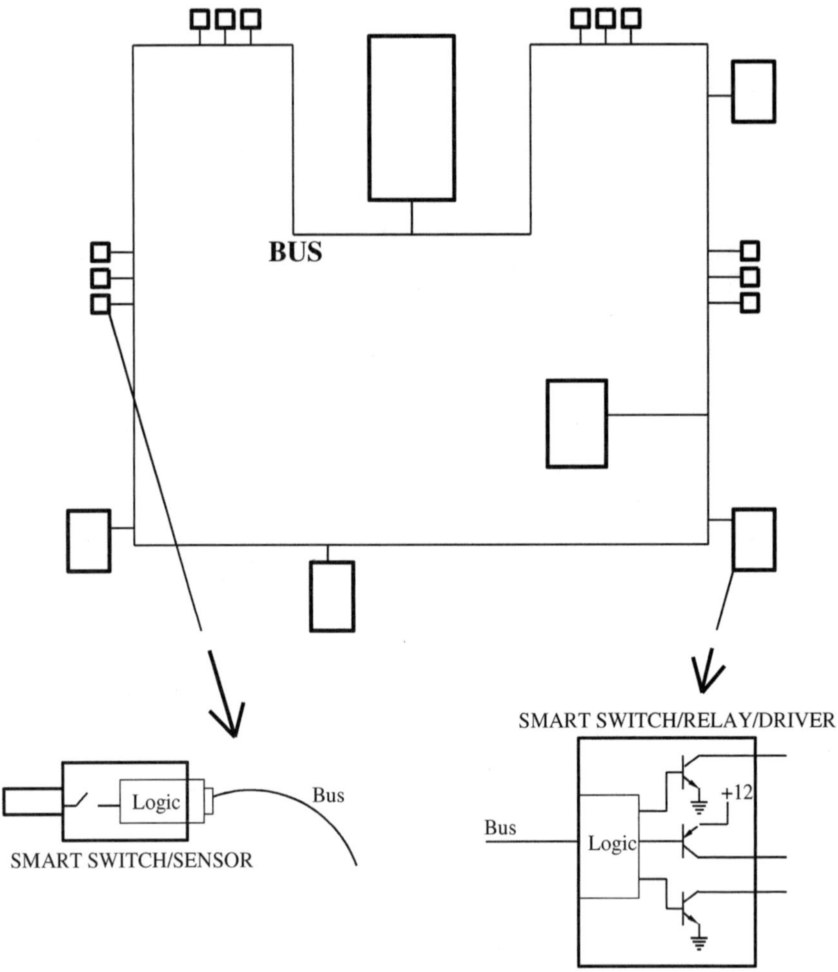

Fig. 20.9 All-vehicle load control multiplexing.

The emphasis seems to be switching from vehicle electronics hardware to software. ECU suppliers will provide generic hardware that is then customized via software written by the vehicle OEM. The following quote sums it up:

"Software. . .describes function in the car. All the uniqueness will be in the software, owned by the carmaker. Hardware will be a commodity item, pounded out by two or three manufacturers in the U.S." [2].

Using generic hardware, the supplier offers the vehicle manufacturer a "black box." The circuitry is designed to a specification without regard to the components used within, except for the multiplex interface. Now, because of the requirement for an interface with a standard serial data link protocol, the vehicle OEMs have opened the door to many ECU vendors. Now, the ECU, say, a transmission controller, is a commodity product.

GM and DaimlerChrysler have publicly admitted interest in pursuing this approach.

Multiplex Competition

Cost, or at least the perception of increased cost, will continually dog the rollout of vehicle multiplexing. [3] This is not the only impediment, however. There are other stumbling blocks and hurdles that have to be dealt with. The first of these is steadily decreasing wire gauge. The choice of most electric/electronic designs for intercontroller connections used to be 18-gauge (1.02-mm) conductor core size insulated wire. Now, 22 gauge (0.643 mm) is typical, and there is a move underway to using 26 gauge (0.404 mm). The change from 22 to 26 gauge saves about 5% in wiring bundle mass. However, only about 65% of the 22 gauge wires may be engineered in 26 gauge because of power constraints and so forth.

Closely related to wire conductor size is the insulation thickness. There are efforts today to decrease the insulation thickness to save mass and size. Conventional "thin-wall" (TW) insulation is being replaced by "ultrathin-wall" (UTW) jacketing. This can give approximately an 11% reduction in the wiring harness mass. The most attractive feature, however, of UTW is the saving in the size of the bundles.

Development of a 36-gauge UTW conductor is underway. A 68% reduction in cross-sectional area is possible compared to earlier wires.

There is an obvious trend toward more fuses in vehicles. This allows lower power circuits, although there has to be more of them. But small UTW wires can be used.

These techniques can be combined (conductor/insulation/fuse) to make a large reduction in the mass and size of wiring bundles. It is unclear how far this can go without creating serious concerns for reliability and serviceability, however.

Proprietary Implementations

Now and in the future, another drawback to the standardization of serial data link protocols is the number of proprietary protocols at vehicle OEMs. Every OEM uses at least one in-house scheme that is not available to other companies (not that they would want it). Some vehicle manufacturers, such as GM and some Asian companies, have no fewer than a dozen in-house protocols. These were all invented for legitimate reasons but create a standardization (and thus serviceability) nightmare.

The big losers to this approach are the IC and scan tool manufacturers. They have to support every protocol if they want the OEM's business. An IC vendor has to design, process, and maintain a large chip set family. Each IC may have a relatively low volume, so prices are high. Scan tool makers have to design-in every possible interface bus, as well as anticipate what will be needed in the future.

A Final Thought

Multiple buses per vehicle have been a reality for some time, primarily because of reuse and carryover of existing networks and protocols. In the near future, however, new functions such as smart connectors, drive-by-wire, and mobile multimedia will demand additional protocols and networks. Despite the best intentions of OEMs, this will continue.

For More Information

For the engineer, manager, and so forth, who works with or is interested in automotive multiplexing, there are a number of sources for further information. It is very difficult to keep up with all the protocols, hardware, manufacturer implementations, and so forth, but the following references can be useful.

Read the following periodicals:

- *EE Times*
- SAE *Automotive Engineering*, especially the April and October issues on electronics
- Ward's *Auto World* magazine
- Chilton's *Automotive Industries* magazine
- "The Hansen Report on Automotive Electronics"

Follow these conferences (attend or read proceedings):

- SAE International Congress and Exposition, usually held in March
- SAE "Convergence," usually in October, every other year
- SAE Symposium on Future Transportation Electronics: Multiplexing and In-Vehicle Networking
- Annual CAN conference (CiA)

References

[1] Allison, N. "Rethinking Multiplex." SAE Paper No. 941650, Society of Automotive Engineers, Warrendale, PA, 1994.

[2] The Hansen Report. June 1995.

[3] The Hansen Report. October 1996.

Appendix A

Glossary of Terms

Acknowledgment:
A type of response that indicates whether a message has been received properly. Acknowledgments can be positive, indicating that the message was received, or negative, indicating that the message was not received.

Application-to-Application Delay Time:
Delay time in a communication network, starting at the request for transmission and terminating at the presentation of the transferred information to the respective application(s).

Arbitration:
The act of two or more nodes transmitting at the same time. Generally, a zero bit "wins" the rights to the bus over a one bit; making a decision relative to a controversy existing at the inputs of the nodes involved.

Arbitration-Based Protocol:
A form of contention-based protocol where contention is evaluated bit-by-bit. Arbitration-based protocols typically require that the physical length of the signal bus be sufficiently short such that the propagation time for signal transmission is significantly less than one information bit time.

Arbitration Field:
Bits within the message frame attributed to each message for controlling the arbitration.

Availability:
The decimal fraction of the time during which a system can perform its required functions. Availability is determined from reliability (mean time between failures) and repair time (mean time to restore) by the relationship

$$A = MTBF/(MTBF + MTTR)$$

To avoid ambiguity, it is possible to define different levels or modes of operation and to determine availability for each level or mode.

Baseband:
The band of frequencies occupied by unmodulated signals. The ratio of the information bandwidth (upper limit minus the lower limit of the frequency band) to the center frequency is typically larger than unity.

Baseband Communications:
A communications method in which the transmitted signal is in its unmodulated form and not changed by modulation.

Baud Rate:
Transfer rate, measured in bauds, which is signal transitions per second.

Bit Rate:
Bits per time during transmission, independent of bit representation.

Bit-by-Bit Contention:
A contention-based arbitration, whereby the contention created by simultaneous access of multiple nodes on the network is resolved bit-by-bit. For example, bits may be represented as dominant and passive on the physical layer; dominant bits override passive bits in case of contention. For this example, the message with a dominant bit in the arbitration field survives without destruction; all others discontinue transmission. This procedure is repeated through all bits of the arbitration field.

Bridge:
A node used to connect two networks that use similar protocols, as differentiated from a gateway.

Broadband Communications:
A communications method in which the transmitted signal is the original signal modulated onto a carrier. The form of modulation can be amplitude, frequency, or phase.

Broadcast Communications:
The transmission of information to more than one receiver, as differentiated from node-to-node communications.

Bus:
A form of data communications channel that does not imply priority among the participants; a physical data channel randomly routed throughout a vehicle upon which each node "taps" at any point and allows bidirectional communication.

CAN:
See Controller Area Network (CAN).

Carrier:
A wave suitable for modulation by an information-bearing signal to be transmitted over a communications medium. For a waveform to be considered to include a carrier, it shall have a separate carrier, as in amplitude and frequency modulation, rather than an inherent carrier, as in pulse-width modulation.

Carrier Sense Multiple Access (CSMA):
An access method in which a node on a multiple node signal bus waits for an idle bus before attempting to transmit.

Carrier Sense Multiple Access with Collision Detection (CSMA/CD):
A type of CSMA method whereby, when a node that is attempting to transmit detects that another node is also attempting to transmit, it delays for a time determined by a predefined contention algorithm.

Carrier Sense Multiple Access with Collision Resolution (CSMA/CR):
A type of CSMA method whereby multiple nodes may begin transmission at the same time, these simultaneous transmissions are arbitrated, and one transmitter ends up retaining control of the link, but the other competing messages are not corrupted.

Centralized Control:
An organization of a control system whereby a central control element exercises control over the remainder of the system (*see also* Master–Slave).

Channel:
A distinct path for data to follow from one node to the next. Channels may be one-way, full-duplex, or half-duplex.

Class 2:
General Motors' version of SAE J1850, or Class B.

Class A:
The SAE class of low-speed data communications; a multiplex system whereby vehicle wiring is reduced by transmitting and receiving multiple signals over the same signal bus between nodes that would have been accomplished by individual wires in a conventionally wired vehicle. The nodes used for multiplexed body wiring typically did not exist in the same or similar form in a totally conventional wired vehicle.

Class B:
The SAE class of medium-speed data communications; a multiplex system, whereby data (e.g., parametric data values) are transferred between nodes to eliminate redundant sensors and other system elements. The nodes in this form of a multiplex system typically already existed as standalone modules in a conventionally wired vehicle.

Class C:
The SAE and GM class of high-speed data communications; a multiplex system, whereby high data rate signals typically associated with real-time control systems, such as engine controls and antilock brakes, are sent over the signal bus to facilitate distributed control and to reduce vehicle wiring further.

Closed System:
A system, consisting of nodes interconnected by a common communications medium (signal bus), which does not permit the easy addition of modules developed by another manufacturer, and temporary connection to other networks, as in an open system.

Coaxial Cable:
A communications medium (signal bus) with two concentric conductors separated by dielectric material(s) resulting in low losses and a relatively constant specific impedance over a wide range of frequencies.

Collision Avoidance Protocol:
A protocol in which a node waits a fixed period of time after the end of the last message or after detecting that another node is also attempting to transmit

before trying to transmit or to transmit again. Each node in a network is assigned a different fixed period of time.

Communication Control Instruction:
The instruction that controls the process of message transfer in a communication system.

Communication Integrity:
A feature of a communication system, whereby the information is transferred uncorrupted and arrives at its destination(s) without modification.

Contention:
A state of the bus in which two or more transmitters are turned on simultaneously to conflicting states.

Contention-Based Protocol:
A protocol or organization of communications data, where the nodes providing the data seek the use of the communications medium (signal bus) and the bus is awarded based on applying a predefined algorithm (typically a priority structure) to the information that the nodes provide.

Controller Area Network (CAN):
A serial data link protocol developed by Bosch in the mid-1980s. Bosch has licensed seven other companies to produce CAN devices.

Data Consistency:
A feature of communications in some multiplex wiring systems, whereby it is determined and ensured that all required recipients of a message have received the message accurately before acting upon it simultaneously. This feature is desirable, for example, to ensure that all four lamps are turned on at once or that all four brakes are energized simultaneously.

Distributed Control:
An organization of a control system, whereby control logic elements are physically located in several different places, as differentiated from centralized organization of control.

Driven Line Length:
The total line length driven by any one node.

Driver:
A solid-state device used to transfer electrical power to the next stage, which may be another driver, an electrical load (power driver), a wire or cable (line, or bus, driver), a display (display driver), and so forth.

Dynamic Priority:
Priority that may be altered during system operation.

Encoding:
A method that tells how to represent information bits in data processing or communication systems.

Error:
The inability of a system to perform properly because of faults.

Error Message:
A special message within a communication network that informs all nodes about an error.

Failure:
The inability of a system to perform properly because of faults.

Fault:
The loss of proper function of a component (hardware or software) that can be either permanent or intermittent.

Fault Tolerance:
The ability of a system to survive a certain number of failures while performing its required functions but, possibly, with some degraded characteristics.

Fiber Optics:
A communications medium (signal bus) consisting of either individual fibers or an assemblage of transparent glass or plastic fiber(s) bundled together parallel to one another. This fiber or bundle of fibers can transmit light along its axis by total internal reflection.

Fiber-Optic Receiver:
An assembly that accomplishes the receive function in fiber-optic communications, typically consisting of a photodetector (either a photodiode or a phototransistor) and a preamplifier.

Fiber-Optic Transmitter:
A unit that accomplishes the driver function in fiber-optic communications and typically consists of a light-emitting diode (LED) and an LED drive circuit. In contrast with the preamplifier of a fiber-optics receiver, the LED drive circuit need not be, and typically is not, packaged with the LED.

Fixed Priority:
The priority preassigned before the start of system operation.

Flexibility:
The ability of a system to function with nodes manufactured by various suppliers.

Flexible System:
A system, consisting of nodes interconnected by a common communications medium (signal bus) according to established standards, to which nodes manufactured by another supplier can be added.

Frequency Division Multiplex Protocol:
A protocol, where the meaning of a bit of information on the signal bus uses the principle of frequency sharing among information channels and the data from each channel are used to modulate sinusoidal signals called subcarriers, so that the resultant signal that represents each channel contains only frequencies in a restricted, narrow frequency range. Multiplex radio transmission, for instance, is the simultaneous transmission of two signals over a common carrier wave.

Full-Duplex:
A data channel, whereby data may go in either direction and at the same time.

Functional Addressing:
Labeling of messages based on their operating code or data content.

Functional Assignment of Wires:
A functional specification for each wire in the physical layer and labeling it with a name.

Gateway:
A system of hardware and software that translates messages in one protocol to another and vice versa; a node used to connect networks that use different protocols, as differentiated from a bridge.

Global Error:
An error in a communication network that is similarly detected in all nodes.

Ground Bus:
The portion of the wiring that serves all multiplex system nodes and provides ground potential and a return path for the current drawn by the node. It is the vehicle chassis in a single-wire bus.

Half-Duplex:
A data channel, whereby data may go in either direction but not at the same time.

Initialization:
Parameterization and eventual configuration of a system during startup.

Internodal Distance:
The signal line length between any two nodes.

Length of Communication Medium:
The maximum distance between any two nodes.

Line Driver:
A solid-state device (driver) used to transfer electrical energy to a wire or cable communications medium (signal bus) that performs the transmit portion of the transceive function.

Line Receiver:
A solid-state device used to receive electrically transmitted signals from a wire or cable communications medium (signal bus) that performs the receive portion of the transceive function.

Link:
Bus; channel; the physical media connecting two or more nodes. Also, a relatively simple communications system that can support communications between two nodes, typically located at the ends of the system.

Listening Mode:
On-mode, receive only; transmission is not allowed.

Local Error:
An error in a communication network that is detected in only some of the nodes.

Master–Slave:
A type of system, whereby one node (a module) acts as a master or central unit and controls the actions of the other nodes, designated as slaves or remote units.

Medium:
The physical manifestation of the link, channel, or bus. It may be an electrical wire or fiber-optic cable.

Message Administration:
The portion of a communication protocol specifying how to handle and to buffer entire messages and respective control bits, for example, which messages shall be transmitted when, or whether messages shall be received and how are they presented to the application.

Message Frame:
A portion of a communication protocol within the message transfer, specifying the arrangement and meaning of bits or bit fields in the sequence of transfer across the transmission medium.

Message Latency:
The time required by a system to access the medium to begin delivering information. Message latency is measured from the time that a node is ready to send specific information to the time that the transmission of this information starts and is ultimately successful. Thus, the total time required to send a desired message successfully is the sum of the message latency and the message transmission time.

Message Rate:
The number of completed message transmissions per unit time.

Message Transfer:
The portion of the protocol that deals with the organization, meaning, and timing of the bits of data. Message transfer deals with which bits must be sent and when they must be sent to transmit a message.

Module:
A subassembly with intelligence, which typically accepts inputs from switches and/or sensors and provides outputs to actuators, lamps, and/or displays. In a multiplex system, a module is one type of node.

Monitoring:
During transmission of a message, the actual physical signal on the transmission line is fed back into the transmitting node for comparison with the transmitted reference signal. Monitoring may be used for error detection that allows safe detection of all global errors in a communication network.

Multimaster:
A system partitioned into several modules, where more than one module may temporarily control the communication network, sending information to or requesting information from other modules.

Multiple Receiver:
The ability to address more than one node with a single message.

Multiplex:
To interleave or simultaneously transmit two or more messages on a single channel.

Multiplex Bus:
The wiring serving all multiplex system nodes and including the signal, power, and ground buses.

Multiplexing:
The process of combining several messages for transmission over the same signal path. There are two widely used methods of multiplexing: time division and frequency division.

Network:
A system that can support communications from three or more nodes.

Network Access Scheme:
A method used to award the communication network to one of the nodes to transmit a message.

Node:
One participant on a bus; contains a protocol handler IC, transceiver, and necessary bus connector; any subassembly of a multiplex system that communicates on the signal bus. In addition to modules, nodes may include other devices that contain the intelligence necessary to support these communications.

Node-to-Node Communications:
The transmission of information to a single receiver, as differentiated from broadcast communications.

Nonreturn to Zero (NRZ):
A data bit format whose voltage or current value (typically voltage) determines the data bit value (typically one or zero).

OBD:
See Onboard Diagnostics II (OBD-II).

Object Layer:
The portion of a communication protocol specifying how to handle and to buffer entire messages and their respective control bits and how to interface the actual application.

Off-Mode:
System behavior when switched off.

Onboard Diagnostics II (OBD-II):
A law begun in California and currently adopted by 12 northeastern states that specifies the inclusion of a certain data link on every new car sold beginning in MY94. The required link must be SAE J1850 or ISO 9141.

One-Way:
A data channel, whereby data may travel in only one direction at all times.

On-Mode:
System behavior in full operation.

Open System:
A system consisting of nodes interconnected by a common communications medium (signal bus) according to established standards that will support temporary connections to manufacturing networks, diagnostics, and other local area networks.

Peer-to-Peer:
A type of network whose individual nodes do not have priorities assigned to govern their access to the bus. All nodes have the same right to access the bus.

Physical Addressing:
Labeling of a message with the physical address of its source and/or destination(s). This is independent of its geographic location, connector pin, and/or wire identification assignments.

Physical Layer:
The properties of the communications medium (signal bus) that can be determined by electrical measurements, such as voltages, currents, impedances, rise times, and so forth.

Power Bus:
The portion of the wiring that serves all multiplex system nodes and provides the electrical power to the nodes and thus the electrical energy used by the node and its associated electrical loads.

Power Driver:
A solid-state device (driver) that can turn electrical loads on and off and requires electrical power significantly in excess of semiconductor logic levels. A power driver can drive actuators, lamps, motors, and so forth. A power driver provides the output function typically provided by a switch in a conventional automotive wiring system but does not provide its input function.

Priority:
The attribute of a message that controls its ranking during arbitration. A high priority increases the probability that a message wins the arbitration process.

Protocol:
A formal set of conventions or rules for exchanging information between nodes, including the procedure for establishing and controlling transmissions on the multiplex signal bus (message administration) and the organization, meaning, and timing of the bits of data (message transfer).

Pulse-Width Modulation:
A data bit format where the width of a pulse of constant voltage or current determines the value (typically one or zero) of the data transmitted.

Receiver:
A device that converts electrical or optical signals used for transmission back to information or data signals.

Recovery from Error Time:
The time delay between detection of an error and restart of regular operation. This may include reinitialization and reconfiguration.

Redundancy:
Added features in a communication and/or data processing system that are not essential for the specified operation but allow the detection of errors or continuation of operation in case of defects.

Reply Time of Acknowledgment:
The time between the end of the source-originated portion of the message and the start of the destination-originated portion of the same message.

Response:
A message or portion of a message initiated by a receiving node as a result of a message transmitted by a different node. A response can be acknowledgment or response data, and it can be appended to the original message (immediate response) or a unique message (separate response).

Response Data:
A response to a message that provides the data (information) requested in the message.

Ring Topology:
A bus topology in which the ends of the bus (line or group of lines) are tied together.

Self-Blocking:
The assignment process of multiple tasks to a single execution instance is called self-blocking if it may get stuck without a chance for recovery.

Signal Bus:
The wire(s) in the portion of the wiring that serve all multiplex system nodes and are dedicated to communications among the nodes.

Sleep Mode:
Node behavior on a low-power consumption standby, waiting to be switched on by a message. This is distinct from an off-mode, where there is no power consumption because the node is disconnected from the power supply.

Star Topology:
Nodes connected by links to a central unit that acts as a central processor or switching point.

Switch:
A mechanically operated device for making, breaking, or changing the connections in an electrical circuit. In a conventional automotive wiring system, a switch serves as both the input device and the output device that provides electrical energy to the load. In a multiplex system, the input device is designated as a switch, and the output device is designated as a driver.

Synchronization:
The procedure to ensure a desired timing for interrelated actions and/or processes.

System Architecture:
The organization of a multiplex system, including, but not necessarily limited to, the location and ranking of logic or decision-making elements, and the types and methods of communications among these elements.

System Elasticity:
The capability of easily adding or deleting nodes and functions that permits expanding or contracting a multiplex wiring system easily, as required. Reprogramming of units not added or deleted should be minimized in an elastic system. This expansion and contraction may be due to model-to-model variations, year-to-year changes, or the desire for new features and accessories.

Time Division Multiplex Protocol:
A protocol where the meaning of a piece of information on the signal bus is determined by its relationship (first, second, third, etc.) to the start of the message or bit stream. In a time division multiplex protocol, data can be interleaved on a bit-by-bit, byte-by-byte, or block-by-block basis.

Token:
The symbol of authority passed between nodes in a token-passing protocol. Possession of this symbol identifies the node currently in control of the medium.

Token-Passing Protocol:
A protocol where a node that has communicated passes the control of the bus, including the right to communicate to another node via a token at the end of the message.

Topology:
The configuration of the interconnected elements of a system.

Total Transmission Latency:
The time required to transfer a message from the transmitting node, measured from the moment it is prepared to send the message, until it is correctly received by the targeted receiver. It may include a retry strategy delay if the initial exchange is not successful.

Transceiver:
An electrical circuit that both transmits (line driver portion) and receives (line receiver portion).

Transfer Delay Time:
The time delay between a request for and the completion of transmission, which is the sum of transmission latency and transmission time.

Transfer Layer:
The portion of a communication protocol that specifies the sequential arrangement and meaning of bits or bit fields within a message. Eventually, the transfer layer may also cover the arbitration scheme, error detection, and handling and strategies for fault confinement.

Transfer Rate:
Information bits per unit time during transmission, equivalent to bit rate.

Transmission Time:
The time for transmitting a message; dependent on message length and transfer rate.

Transmitter:
A device that converts information or data signals to electrical or optical signals so that these signals can be sent over a communications medium (signal bus).

Twisted Pair:
A cable composed of two insulated conductors twisted about one another.

Variable Pulse-Width Modulation (VPWM):
A method of using both bus state and pulse width to encode bit information. This encoding technique is used to reduce the number of bus transitions for a given bit rate. One embodiment would define a one ("1") as a dominant short pulse or a passive long pulse, whereas a zero ("0") would be defined as a long dominant pulse or a short passive pulse. Because a message is composed of random 1's and 0's, general byte or message times cannot be predicted in advance.

Appendix B

List of Acronyms

4WD	Four-Wheel Drive
4WS	Four-Wheel Steering
AAMA	Automobile Manufacturers Association
ABS	Antilock Brake System
ABUS	Asynchrone Bitserielle Unversalschnittstelle; *also* Automobile BUS
ACP	Audio Control Protocol
ALDL	Assembly Line Diagnostics Link
AMI-C	Automotive Multimedia Interface Collaboration
ANSI	American National Standards Institute
ARCNET	Attached Resource Computer Network
ASI	Actuator Sensor Interface
ASIC	Application-Specific Integrated Circuit
BB	Building Block
BBTAG	Broadband Technical Advisory Group
BBWA	Broadband Wireless Access
BCI	Bulk Current Injection
BDLC	Byte Data Link Controller
BEAN	Body Electronic Area Network
BIT	Built-In Test
CAD	Computer Aided Design
CAM	Computer Aided Manufacturing
CAN	Controller Area Network
CARB	California Air Resources Board
CCD	DaimlerChrysler Collision Detection; *also* C^2D
CiA	CAN in Automation
CMOS	Complementary Metal Oxide Semiconductor
CPU	Central Processor Unit
CRC	Cyclic Redundancy Check

CRT	Cathode Ray Tube
CSIC	Customer-Specific Integrated Circuit
CSMA	Carrier Sense Multiple Access
CSMA/CD	Carrier Sense Multiple Access with Collision Detection
CSMA/CR	Carrier Sense Multiple Access with Collision Resolution
CTIA	Cellular Telecommunications Industry Association
DIP	Dual In-Line Package
DLC	Data Link Controller; *also* Data Link Connector; *also* Data Length Count
DLCM	Data Link Controller Module
DLCP	Data Link Controller Parallel
DLCS	Data Link Controller Serial
DLP	Data Links Partnership; *see* HSSDCPPWG
DTC	Diagnostic Trouble Code
E&C	Entertainment and Comfort
ECM	Engine Control Module
ECU	Electronic Control Unit
EGR	Exhaust Gas Recirculation
EMC	Electromagnetic Compatibility
EMI	Electromagnetic Interference
EOD	End of Data
EOF	End of Frame
EPA	U.S. Environmental Protection Agency
ESD	Electrostatic Discharge
EWCAP	Electrical Wiring Component Application Partnership
FCC	Federal Communications Commission
FFT	Fast Fourier Transform
FIFO	First-In First-Out
FOTAG	Fiber Optics Technical Advisory Group
GPIB	General Purpose Interface Bus
GPS	Global Positioning System
HART	Highway Addressable Remote Transducer
HBCC	Hosted Bus Communications Controller
HDTV	High-Definition Digital Television

HILI	High-Level Interface
HSSDCPPWG	High-Speed Serial Data Communications Partnership Protocol Work Group; *see* DLP
HVAC	Heating, Ventilation, and Air Conditioning
HVCMOS	High-Voltage Complementary Metal Oxide Semiconductor
IC	Integrated Circuit
IDB	Intelligent Transportation Systems Data Bus (Committee)
IEEE	Institute of Electrical and Electronic Engineers
IFR	In-Frame Response
IMS	Inter-Message Separation
I/O	Input/Output
IP	Instrument Panel
ISLAN	Integrated Services Local Area Network
ISO	International Organization for Standardization
ITS	Intelligent Transportation Systems
JRP	Joint Research Partnership
JSAE	Japanese Society of Automotive Engineers
kb/s	Kilobits per Second
kB/s	Kilobytes per Second
KWP 2000	Keyword Protocol 2000
LBCC	Low-Cost Bus Communication Controller
LCU	Local Control Unit
LDT	Light-Duty Truck
LDV	Light-Duty Vehicle
LED	Light-Emitting Diode
LIN	Local Interconnect Network
LLC	Logical Link Control
LSb	Least Significant Bit
LSB	Least Significant Byte
MAC	Medium Access Control
MAN	Metropolitan Area Network
MDLC	Message Data Link Controller
MFM	Modified Frequency Modulation
MICS	Mitsubishi Intelligent Cockpit System

MID	Message Identifier
MIL	Malfunction Illumination Lamp
MISO	Master In Slave Out
MOS	Metal Oxide Semiconductor
MOSI	Master Out Slave In
MOST	Media-Oriented Systems Transport
MSb	Most Significant Bit
MSB	Most Significant Byte
MY	Model Year
N/A	Not Allowed; Not Applicable
NB	Normalization Bit
NRZ	Nonreturn to Zero
OBD	Onboard Diagnostics
ODVA	Open DeviceNet Vendor Association
OEM	Original Equipment Manufacturer
OSEK/VDX	Offene Systeme und deren Schnittsellen fur die Elektronik im Kraftfahrzeug/Vehicle Distributed Executive; *also* Open Systems and Interfaces for Distributed Electronics in Cars
OSI	Open Systems Interconnect
PAL	Programmable Array Logic
PALMNET	Protocol for Automotive Local Area Network
PCI	Programmable Controller Interface
PCM	Powertrain Control Module
PDH	Pleisiochronous Digital Heirarchy
PID	Parameter Identifier
PIN	P-Doped Intrinsic and N-Doped
PLC	Programmable Logic Control
PLCC	Plastic Leaded Chip Carrier
PNGV	Partnership for a New Generation of Vehicles
POF	Plastic Optical Fiber
PROM	Programmable Read-Only Memory
PWM	Pulse-Width Modulation
QFD	Quality Function Deployment
QFP	Quad Flat Pack

RF	Radio Frequency
RFI	Radio-Frequency Interference
RFIFO	Receive First-In First-Out
ROM	Read Only Memory
RPRSG	Resilient Packet Ring Study Group
RZ	Return to Zero
SAE	Society of Automotive Engineers (SAE International)
SAVA	Standard Army Vetronics Architecture
SCI	Serial Communications Interface
SCP	Standard Corporate Protocol
SILS	Standard for Interoperable Local Area Network (LAN) Security
SIR	Supplemental Inflatable Restraint
SJW	Synchronization Jump Width
SLIO	Serial Link Input/Output
SMI	Serial Multiplex Interface
SOF	Start of Frame
SPI	Serial Peripheral Interface
STP	Shielded Twisted Pair
SWC	Single-Wire CAN
TCM	Tansmission Control Module
TCS	Traction Control System
TFIFO	Transmit First-In First-Out
TMC	Truck Maintenance Council
TP	Twisted Pair
TPU	Time Processor Unit
TS	"Token–Slot" Protocol
TTCAN	Time-Triggered Controller Area Network
TTP	Time-Triggered Protocol
TW	Thin-Wall [Insulation]
UART	Universal Asynchronous Receiver/Transmitter
USB	Universal Serial Bus
USCAR	United States Council for Automotive Research
UTP	Unshielded Twisted Pair
UTW	Ultrathin-Wall [Jacketing]
UWB	Ultrawideband

VAN	Vehicle Area Network
VCM	Vehicle Control Module
VHDL	VHSIC Hardware Description Language
VPW	Variable Pulse-Width
VPWM	Variable Pulse-Width Modulation
WLAN	Wireless Local Area Network
WPAN	Wireless Personal Area Network
XCVR	Transceiver

Appendix C

Bus Comparison Tables

BUS PROTOCOL COMPARISON									
	BUS TYPE								
FEATURE	SINGLE-WIRE CAN	CAN 2.0 ISO 11898 ISO 11519-2 ISO 11992 J2284 J1939	J1850 ISO 11519-4			CCD	ACP	BEAN	IP-BUS
AFFILIATION	SAE/ISO	BOSCH/SAE/ISO	GM	FORD	DAIMLER-CHRYSLER	DAIMLER-CHRYSLER	FORD	TOYOTA	PIONEER
APPLICATION	DIAGNOSTICS	CONTROL & DIAGNOSTICS	GENERAL & DIAGNOSTICS	GENERAL & DIAGNOSTICS	GENERAL & DIAGNOSTICS	GENERAL & DIAGNOSTICS	AUDIO CONTROL	BODY CONTROL & DIAGNOSTICS	ENTERTAINMENT
TRANSMISSION MEDIUM	SINGLE WIRE	TWISTED PAIR	SINGLE WIRE	TWISTED PAIR	SINGLE WIRE	SINGLE WIRE	TWISTED PAIR	SINGLE WIRE	TWISTED PAIR
BIT ENCODING	NRZ-5 MSb FIRST	NRZ-5 MSb FIRST	VPW MSb FIRST	PWM MSb FIRST	VPW MSb FIRST	NRZ	NRZ	NRZ	PWM
MEDIA ACCESS	CONTENTION	CONTENTION	CONTENTION	CONTENTION	CONTENTION	MASTER/ SLAVE	MASTER/ SLAVE	CONTENTION	CONTENTION
ERROR DETECTION	CRC	CRC	CRC	CRC	CRC	8-BIT CS	8-BIT CS	8- BIT CRC	PARITY
HEADER LENGTH	11 BITS	11 OR 29 BITS	32 BITS	32 BITS	8 BITS	8 BITS	12–24 BITS	25 BITS	45 BITS
DATA FIELD LENGTH	0–8 BYTES	0–8 BYTES 11- OR 29-BIT ID	0–8 BYTES	0–8 BYTES	0–10 BYTE	5 BYTES	6–12 BYTES	1–11 BYTES	32 BYTES
MESSAGE OVERHEAD	9.9%	9.9–22%	33.3%	33.3%	8.3%	16.7%	25%	28%	17.5%
IN-MESSAGE RESPONSE	NO	NO	OPTIONAL NORMALLY NO	OPTIONAL NORMALLY YES	OPTIONAL NORMALLY YES	NO	NO	NO	NO
BIT RATE	20.8–125 kb/s	1 Mb/s	10.4 kb/s	41.6 kb/s	10.4 kb/s	7812.5 b/s	9600 b/s	10 kb/s	26 kb/s
MAXIMUM BUS LENGTH	30 METERS	NOT SPECIFIED 40 (TYPICAL)	35 METERS (5 M OFFBOARD)	35 METERS (5 M OFFBOARD)	35 METERS (5 M OFFBOARD)	NOT SPECIFIED	40 METERS	NOT SPECIFIED	120 METERS
MAXIMUM NODES	16	32 (TYPICAL)	32	32	32	6	20	20	20
µ NEEDED?	YES	YES	YES	YES	YES	YES	YES	YES	YES
SLEEP/WAKE-UP	YES	NO	YES	NO	NO	NO	NO	NO	NO
H/W AVAILABLE?	NO	YES	YES	YES	YES	YES	YES	YES (?)	YES
COST	LOW	MEDIUM	LOW	LOW	LOW	LOW	LOW	LOW	LOW

BUS PROTOCOL COMPARISON (cont.)									
	BUS TYPE								
FEATURE	SLIO CAN	SINEBUS	SAFETYBUS	ISO/DIS 9141 ISO/DIS 9141-2 KEYWORD XX	ABUS	VAN ISO 11519-3	MI-BUS	I^2C	TTP
AFFILIATION	PHILIPS	DELPHI	DELPHI	WORLD	VOLKSWAGEN	RENAULT	MOTOROLA	PHILIPS	U-VIENNA
APPLICATION	CONTROL & DIAGNOSTICS	GENERAL	CONTROL & DIAGNOSTICS	DIAGNOSTICS ONLY	CONTROL & DIAGNOSTICS	CONTROL & DIAGNOSTICS	CONTROL		CONTROL & DIAGNOSTICS
TRANSMISSION MEDIUM	TWISTED PAIR	SINGLE WIRE	TWISTED PAIR	SINGLE WIRE	SINGLE WIRE	TWISTED PAIR	SINGLE WIRE	TWISTED PAIR	NOT SPECIFIED
BIT ENCODING	NRZ-5 MSB FIRST	SAM	RTZ (10 LEVELS)	NRZ (START, 7D, P, STOP) LSb FIRST	NRZ LSb FIRST	ENHANCED MANCHESTER	MANCHESTER	AM	NOT SPECIFIED
MEDIA ACCESS	CONTENTION	MASTER/ SLAVE	CONTENTION	TESTER/ SLAVE	CONTENTION	CONTENTION	MASTER/ SLAVE		TIME DIVISION
ERROR DETECTION	CRC	NONE REQUIRED	8-BIT CRC	PARITY (ODD)	BIT ONLY	CRC		ACK BIT	16-BIT CRC
HEADER LENGTH	11 BITS	2 BITS	1 BYTE	NOT SPECIFIED	12 BITS	12 BITS	1 BYTE		1 BYTE
DATA FIELD LENGTH	0–8 BYTES	10–18 BITS	24–39 BITS	NOT SPECIFIED	2 BYTES	0–28 BYTES	1 BYTE		16 BYTES
MESSAGE OVERHEAD	9.9%	75%	20–30%	Variable	48%	41.3%	50%	45%	18.75%
IN-MESSAGE ACK.	NO	NO	NO	NO	NO	YES	NO		YES
BIT RATE	20–125 kb/s	66.6 KHz	500 kb/s	<10.4 kb/s	500 kb/s	125-250 kb/s	20 kb/s	1–100 kb/s	NOT SPECIFIED
MAXIMUM BUS LENGTH	40 METERS	10 METERS	NOT SPECIFIED	LIMITED BY TOTAL IMPEDANCE TO GROUND	NOT SPECIFIED (30 M TYPCIAL)	20 METERS	NOT SPECIFIED	NOT SPECIFIED	NOT SPECIFIED
MAXIMUM NODES	16		64	LIMITED BY TOTAL IMPEDANCE TO GROUND	NOT SPECIFIED 32 (TYPICAL)	16	16	NOT SPECIFIED	NOT SPECIFIED
μ NEEDED?	NO	NO	?	YES	YES	YES	NO		YES
SLEEP/WAKE-UP	YES	NO	NO	NO	NO	NO			NO
H/W AVAILABLE?	YES	NO	NO	YES	YES	YES	YES		NO
COST	LOW	LOW	LOW	LOW	MEDIUM	MEDIUM			UNKNOWN

BUS PROTOCOL COMPARISON (cont.)

FEATURE	**BUS TYPE** PALMNET	J1587	J1922	D^2B Copper	D^2B Optical	MML	USB	IEEE 1394	SCI
AFFILIATION	MAZDA	SAE T&B	SAE T&B	PHILIPS	PHILIPS	DELPHI		IEEE	
APPLICATION	CONTROL	GENERAL & DIAGNOSTICS	CONTROL	STREAM DATA & CONTROL	STREAM DATA & CONTROL	STREAM DATA & CONTROL	PC DEVICES	PC DEVICES	PC DEVICES
TRANSMISSION MEDIUM	TWISTED PAIR	TWISTED PAIR	TWISTED PAIR	TWISTED PAIR	OPTICAL	OPTICAL FIBER	SHIELDED TWISTED PAIR	SHIELDED TWISTED PAIR	3 WIRES
BIT ENCODING	NRZ	NRZ	NRZ	PWM	BIPHASE	NRZ	NRZ	NRZ	NRZ
MEDIA ACCESS	CONTENTION	MASTER/ SLAVE	MASTER/ SLAVE	MASTER/ SLAVE	MASTER/ SLAVE	MASTER/ SLAVE	CONTENTION	CONTENTION	FULL DUPLEX
ERROR DETECTION	8-BIT CRC	CHECKSUM	CHECKSUM	PARITY	CRC	CORRECTING (OPTIONAL)	CRC	CRC	
HEADER LENGTH	4 BYTES	1 BYTE	1 BYTE			1 BYTE			1 BYTE
DATA FIELD LENGTH	4 BYTES	1–19 BYTES	1–20 BYTES			1–200+ BYTES			1–200+ BYTES
MESSAGE OVERHEAD	50%	VARIABLE	VARIABLE			5–10%	25%	25–30%	VARIABLE
IN-MESSAGE ACK.	YES	NO	NO			NO			NO
BIT RATE	20–1000 kb/s	9600 b/s	9600 b/s	29.8 kb/s	12 Mb/s	110 Mb/s	12 Mb/s	98–393 Mb/s	35–131 kb/s
MAXIMUM BUS LENGTH	20 METERS	40 METERS	40 METERS	150 METERS	NO LIMIT	10 METERS		72 METERS	50 FEET
MAXIMUM NODES	NOT SPECIFIED	20	4	50	24	16	127	16	16–32
μ NEEDED?	YES	YES	YES	YES	YES	YES	YES	YES	YES
SLEEP/WAKE-UP		NO	NO			YES	NO	NO	YES
H/W AVAILABLE?	MAYBE	YES	YES	YES	YES	NO	YES	YES	YES
COST		LOW	LOW	HIGH	HIGH	HIGH	MEDIUM	MEDIUM	LOW

BUS PROTOCOL COMPARISON (cont.)									
	BUS TYPE								
FEATURE	SPI	ST BUS	SBI	RS 232 EIA RS562	RS 422 EIA 485 RS 485	UART(ALDL)	E&C	KDD	TSN
AFFILIATION			DELPHI	EIA	EIA	GM	GM	DELPHI	GM
APPLICATION	μ PERIPHERALS		GENERAL	PC DEVICES	PC DEVICES	GENERAL & DIAGNOSTICS	GENERAL	DISPLAY & KEYBOARD CONTROL	REAL-TIME CONTROL
TRANSMISSION MEDIUM	4 WIRES	2 WIRES	SINGLE WIRE	3 WIRES	2 WIRES	SINGLE WIRE	SINGLE WIRE	SINGLE WIRE	NOT SPECIFIED
BIT ENCODING	AM	AM	NRZ	NRZ	AM	NRZ	PWM		NRZ-5
MEDIA ACCESS	MASTER/ SLAVE	MASTER/ SLAVE	MASTER/ SLAVE	MASTER/ SLAVE	MASTER/ SLAVE	MASTER/ SLAVE	CONTENTION	MASTER/ SLAVE	TOKEN SLOT
ERROR DETECTION	NOT SPECIFIED		CHECKSUM			8-BIT CS	PARITY	NONE	16-BIT CRC
HEADER LENGTH	VARIABLE		1 BYTE			16 BITS	11–12 BITS	VARIABLE	16 BITS
DATA FIELD LENGTH	19–29 BITS	3 BITS	3–4 BYTES			0–85 BYTES	1–8 BITS	VARIABLE	0–256 BYTES
MESSAGE OVERHEAD	VARIABLE		VARIABLE	VARIABLE	VARIABLE	VARIABLE	VARIABLE	VARIABLE	VARIABLE
IN-MESSAGE ACK.	NO		NO	NO	NO	NO	NO	NO	YES
BIT RATE	< 4.2 Mb/s	1 Mb/s	6.67 kb/s	20 kb/s	10 Mb/s	8192 b/s	1000 b/s	1000 b/s	Up to 2 Mb/s+
MAXIMUM BUS LENGTH	< 10 METERS	4 FEET	8 METERS	15 METERS	1200 METERS	NOT SPECIFIED	20 METERS	NOT SPECIFIED	NOT SPECIFIED
MAXIMUM NODES	NOT SPECIFIED		32	32	32	10	10	2	32
μ NEEDED?	YES	YES	YES	YES	YES	YES	YES	1 PER BUS	YES
SLEEP/WAKE-UP	NO	NO	YES	YES	YES	NO	NO	NO	NO
H/W AVAILABLE?	YES	YES	YES	YES	YES	YES	YES	YES	YES
COST	LOW	LOW	LOW	LOW	LOW	LOW	VERY LOW		HIGH

BUS PROTOCOL COMPARISON (cont.)

	BUS TYPE								
FEATURE	SAE J1708	SAE J1939	POWER LINE CARRIER—SPREAD SPECTRUM	POWER LINE CARRIER—QPSK	QWIK	AVC LAN	J BUS	C BUS	SC Bus
AFFILIATION	TMC–ATA	TMC–ATA	Intellon	YAMAR	QED	Toyota	GM T & B	Clarion	RDS/Delphi
APPLICATION	CONTROL & DIAGNOSTICS	CONTROL & DIAGNOSTICS	CONTROL & DIAGNOSTICS	ENTERTAIN-MENT ONLY	BODY CONTROL	CONTROL	CONTROL	CONTROL	GENERAL
TRANSMISSION MEDIUM	TWISTED PAIR	TWISTED PAIR	POWER LINE	POWER LINE	1 WIRE				1 WIRE
BIT ENCODING	NRZ	NRZ-5 MSb first	PHASE REVERSING KEY (PRK)	QUAD PHASE SHIFT KEYING (QPSK)	NRZ		PULSE CODE MODULATION (PCM)	PULSE CODE MODULATION (PCM)	NRZ
MEDIA ACCESS	MASTER/ SLAVE	CONTENTION	MASTER/SLAVE OR CONTENTION		ISOCHRONOUS				MASTER/ SLAVE
ERROR DETECTION	8-BIT CS	CRC	8-BIT CS		NONE	EVEN 1's PARITY	8-BIT CS	ECHO BACK EACH BYTE	ACK BYTE
HEADER LENGTH	16 BITS	11 OR 29 BITS	16 BITS		NONE				3 BITS
DATA FIELD LENGTH		0–8 BYTES 11 OR 29-BIT ID			8 BITS/NODE				1–2 BYTES
MESSAGE OVERHEAD	VARIABLE	9.9–22%	VARIABLE		NONE				VARIABLE
IN-MESS ACK?.	NO	NO			NO				YES
BIT RATE	9600	1 Mb/s	9600	6.8 Mb/s	25 kb/s	3.9–26 kb/s	9600/19.2 kb/s	262 kb/s	10 kb/s
MAXIMUM BUS LENGTH	NOT SPECIFIED	NOT SPECIFIED 40 (TYPICAL)	NOT SPECIFIED	NOT SPECIFIED	NOT SPECIFIED		40 METERS		8 METERS
MAXIMUM NODES		NOT SPECIFIED 32 (TYPICAL)		16+	32	15	20		8
μ NEEDED?	YES	YES	YES	YES	NO	YES	YES		YES
SLEEP/WAKE-UP	NO	NO	NO	NO	NO				YES
H/W AVAILABLE?	YES	YES	YES	YES	YES	YES			YES
COST	MEDIUM	MEDIUM	MEDIUM	MEDIUM	LOW	MEDIUM			LOW

BUS PROTOCOL COMPARISON (cont.)										
	BUS TYPE									
FEATURE	IDB-I (ITSDATA BUS)	IDB-II (ITS DATA BUS)	MOST (D^2B)	KEYWORD XX						
AFFILIATION	SAE	SAE	Philips	Various						
APPLICATION	AFTERMARKET ENTERTAIN-MENT	AFTERMARKET ENTERTAIN-MENT	STREAM DATA & CONTROL	DIAGNOSTICS						
TRANSMISSION MEDIUM	1-WIRE	TBD	OPTICAL	SINGLE WIRE						
BIT ENCODING	NRZ	TBD	BIPHASE	NRZ						
MEDIA ACCESS	TBD	TBD	MASTER/ SLAVE	MASTER/ SLAVE						
ERROR DETECTION			CRC	X-BIT CS						
HEADER LENGTH				16 BITS						
DATA FIELD LENGTH				0–85 BYTES						
MESSAGE OVERHEAD				VARIABLE						
IN-MESSAGE ACK.				NO						
BIT RATE	115.2 kb/s	HIGH	25 Mb/s	5 b/s–10.4 kb/s						
MAXIMUM BUS LENGTH	TBD	TBD	INFINITE	NOT SPECIFIED						
MAXIMUM NODES	16		24	10						
μ NEEDED?	YES	YES	YES	YES						
SLEEP/WAKE-UP	YES	YES		NO						
H/W AVAILABLE?	NO	NO	YES	YES						
COST	LOW	HIGH	HIGH	LOW						

OPEN FIELD BUS PROTOCOL COMPARISON									
	BUS NAME								
FEATURE	FOUNDATION FIELDBUS (IEC 1158 ISA S50)	PROFIBUS (EN 50170 DIN 19245)	CONTROLNET (EN 50254)	MODBUS	INTERBUS	ETHERNET (10Base5 IEEE802.3 ISO8802.3)	WORLDFIP (EN 50170)	LONWORKS	HART
AFFILIATION	FIELDBUS FOUNDATION	PROFIBUS TRADE ORGANIZATION			INTERBUS CLUB		WORLDFIP EUROPE	LONMARK INTEROPER-ABILITY ASSOCIATION	HART COMMUNICATION FOUNDATION
APPLICATION	PROCESS CONTROL INSTRUMENTA-TION AUTOMATION	PROCESS CONTROL INSTRUMENTA-TION AUTOMATION	PROCESS CONTROL AUTOMATION	PROCESS CONTROL AUTOMATION	PROCESS CONTROL AUTOMATION	PROCESS CONTROL	PROCESS CONTROL INSTRUMENTA-TION AUTOMATION	PROCESS CONTROL AUTOMATION	INSTRUMEN-TATION
TRANSMISSION MEDIUM	OPTICAL FIBER RF, TP	RS-485 (TP)	OPTICAL FIBER COAX	TP	OPTICAL FIBER TP	COAX	OPTICAL FIBER RF, TP	OPTICAL FIBER RF, TP, COAX	TP
BIT ENCODING									
MEDIA ACCESS	TOKEN PASSING	TOKEN PASSING	CONTENTION	TOKEN PASSING	TOTAL FRAME (?)	CSMA/CD	BUS-ARBITER (?)		MASTER/SLAVE
ERROR DETECTION									
DETERMINISTIC?	YES	NO	YES	NO	YES	NO	YES	NO	NO
HEADER LENGTH									
DATA FIELD LENGTH									
MESS. OVERHEAD									
IN-MESSAGE ACK.									
BIT RATE	31.25K–2.5M	31.25K–12M	5 Mb/s	1.2–115.2 K	500 kb/s	10 Mb/s	31.25K–2.5 M	300K–1.25 M	1200 b/s
MAXIMUM BUS LENGTH									
MAXIMUM NODES	240/SEGMENT 2^{16}/NETWORK	256	99	247	512	400/SEGMENT	256	32,768/DOMAIN	15/LOOP
μ NEEDED?									
SLEEP/WAKE-UP									
H/W AVAILABLE?	PLANNED	YES	YES	NO	YES	YES	YES	YES	YES
COST	~$20/NODE	~$20/NODE						VARIES	
CONTACT	www.fieldbus.org	www.profibus.com			www.interbusclub.com		www.worldfip.org	www.lonmark.org	www.hartcomm.org

OPEN FIELD BUS PROTOCOL COMPARISON (cont.)										
	BUS NAME									
FEATURE	SDS	DEVICENETt	AS-INTERFACE	SERCOS	SERIPLEX					
AFFILIATION	HONEYWELL	ODVA (ALLEN-BRADLEY)								
APPLICATION	PROCESS CONTROL AUTOMATION	PROCESS CONTROL AUTOMATION	PROCESS CONTROL AUTOMATION							
TRANSMISSION MEDIUM	FIBER OPTIC STP	FIBER OPTIC STP	TP	FIBER OPTIC	FLAT CABLE (4 WIRES)					
BIT ENCODING	NRZ-5	NRZ-5								
MEDIA ACCESS	CONTENTION	CONTENTION	MASTER/ SLAVE							
ERROR DETECTION	15-BIT CRC	15-BIT CRC								
DETERMINISTIC?	NO	NO	YES							
HEADER LENGTH										
DATA FIELD LENGTH	8 BYTES	8 BYTES								
MESS. OVERHEAD										
IN-MESSAGE ACK.										
BIT RATE	UP TO 1 Mb/s	UP TO 1 Mb/s	167 kb/s							
MAXIMUM BUS LENGTH										
MAXIMUM NODES	2^{11} to 2^{29}	2^{11} to 2^{29}	31							
μ NEEDED?										
SLEEP/WAKE-UP										
H/W AVAILABLE?	PLANNED	YES	YES							
COST	~$20/NODE	~$10–$20/NODE	~$5/NODE	~$20/NODE	~$10/NODE					
CONTACT	www.honeywell.com/sensing/prodinfo/sds/	www.devicenet.org	www.as-interface.com							

Appendix D

List of SAE and ISO Committees

SAE:

J1587	T&B UART General Purpose Diagnostic Message Format
J1699	J1850 Compliance Procedure
J1708	T&B UART Physical Layer
J1850	Class B Communications
J1922	T&B UART Powertrain Diagnostic Message Format
J1939	T&B High-Speed CAN
J1962	Diagnostic Interface Connector
J1978	Scan Tool Requirements
J1979	CARB/EPA Messages
J2012	Diagnostic Codes
J2057	Class A Multiplexing Definitions
J2178	Basic Diagnostic Message Format
J2190	Service Diagnostic Messages
J2201	Scan Tool Physical Interface
J2205	Optional Expanded Diagnostic Protocol
J2284	Passenger Car High-Speed CAN
J2300	Compliance Procedure for Scan Tools
J2411	Single Wire CAN
J2480	CAN for OBD-III
J2602	LIN

ISO:

9141	General Purpose UART Diagnostic
9141-2	OBD-II UART Diagnostic
9141-3	OBD-II UART Diagnostic Verification Procedures
11519-2	Low-Speed CAN
11898	High-Speed CAN
11992	T&B Tractor-Trailer CAN Interface

14229	General Diagnostic Procedures
14230	Keyword 2000
15031	CAN Emissions Diagnostics
15765	CAN General Diagnostics
16845	CAN Conformance

Index

About the Author

Christopher (Chris) A. Lupini has a Bachelor of Science degree in Computer Engineering from the University of Michigan and a Master of Science degree in Electrical Engineering from Purdue University. He is a registered Professional Engineer and an ASE certified Master Automotive Technician. He has worked in the data communications industry for 15 years. Currently, he is a staff research engineer for Delphi Delco Electronics Systems and has also been consulting and teaching for 10 years.

Mr. Lupini has spent most of his time assisting in designing and developing the General Motors Class 2 protocol, as well as the specification and testing of numerous J1850 and CAN ICs. He is the lead engineer for the Delphi Serial Data Center of Expertise. He has authored a dozen technical papers and articles plus a textbook, and he has one patent.